TOPOGRAPHICAL WRITERS
IN SOUTH-WEST ENGLAND

EXETER STUDIES IN HISTORY
General Editors: Jonathan Barry, Tim Rees
and T.P. Wiseman

Also in the Exeter Studies in History series:

The Inheritance of Historiography, 350–900,
edited by Christopher Holdsworth and T.P. Wiseman (1986)

Security and Defence in South West England before 1800,
edited by Robert Higham (1987)

Landscape and Townscape in the South West,
edited by Robert Higham (1989)

The Saxon Shore: A handbook,
edited by Valerie Maxfield (1989)

Landscape and Settlement in Britain AD 400–1066,
edited by Della Hooke and Simon Burnell (1995)

*From Deliverance to Destruction: Rebellion and
Civil War in an English City*,
by Mark Stoyle (1996)

TOPOGRAPHICAL WRITERS IN SOUTH-WEST ENGLAND

edited by
Mark Brayshay

UNIVERSITY
of
EXETER
PRESS

First published in 1996 by
University of Exeter Press
Reed Hall, Streatham Drive
Exeter, Devon EX4 4QR
UK

British Library Cataloguing in Publication Data
A catalogue record of this book is
available from the British Library

ISBN 0 85989 424 X

Typeset in 11 on 13pt Goudy Old Style
by GreenShires Icon, Exeter
Printed and bound in Great Britain
by Short Run Press Ltd, Exeter.

CONTENTS

Illustrations

Cover illustrations:
The map showing part of the South-West of England was first published as a frontispiece in William Marshall, *The Rural Economy of the West of England* (London, 1796). The original map is ascribed to 'H. Mutlow and C. Russell Co. London'.

The portrait is of Richard Carew (1555–1620) of Antony in Cornwall, author of *The Survey of Cornwall* (London, 1602). The original of this picture, which was painted in 1586 when Carew was only 32, still hangs in the Hall of Antony House and shows him holding a book on which is inscribed: *Invita morte vita*. Behind him there is a device of a hammer striking an anvil within which there is another inscription: *Chi verace durera* (He who is trustworthy will endure). [*Reproduced by kind permission of Sir Richard Carew Pole*].

Notes on Contributors

Joseph Bettey, MA PhD FSA FRHistS was formerly Reader in Local History at the University of Bristol, and is author of numerous books and papers on West-Country history including *Wessex from AD 1000* (1986). Particular research interests have included agriculture, rural society and church life in Wiltshire, Somerset and Dorset during the sixteenth and seventeenth centuries. He is general editor of the Bristol Record Society.

Mark Brayshay, BA PhD is a principal lecturer in Geography at the University of Plymouth where he teaches special courses in historical geography. He read Geography at the University of Exeter and his doctoral thesis focused on mid-Victorian west Cornwall. He is author of many papers which focus on his research interests including early modern road travel and communications, nineteenth-century emigration, Victorian model cottages and public health improvement.

John Chandler, BA PhD is a freelance historical researcher based in Wiltshire. He was brought up in Exmouth, studied Classics at the University of Bristol, and worked as local studies librarian for Wiltshire County Council until 1988. He is author of several books on Wiltshire history, and of an edition of John Leland's *Itinerary*. He is also general editor of the Wiltshire Record Society.

Robert Dunning, BA PhD FSA has been editor of the *Victoria History of Somerset* since 1967. He read History at the University of Bristol and his doctoral thesis was on the administration of the diocese of Bath and Wells in the Later Middle Ages. He is author of books and articles on Somerset history and is honorary general editor of the Somerset Record Society.

Ian Maxted, MA has worked for Devon County Council's local studies libraries in Exeter since 1977 and, prior to that, he was assistant keeper in Guildhall Library, London. He has published a number of books and articles on book history and has compiled various bibliographies including the Devon section of the *Bibliography of British Newspapers* and the *Devon Bibliography* annually since 1985.

Malcolm Todd, BA DLitt DipArchaeol FSA has been Professor of Archaeology at the University of Exeter since 1979. His scholarly interests centre on the archaeology of the Roman Empire and the Migration Period in western Europe. His publications include *The Northern Barbarians* (1975), *Roman Britain* (1981) and *The Early Germans* (1992), and he is currently working on a large study of Late Antiquity. Among his fieldwork projects is a study of the ancient sites of mining in the Mendips. He is also interested in the place of antiquarian and archaeological studies in the history of ideas.

Sarah Wilmot, MA PhD is a research fellow at the University of Cambridge. Since completing her PhD on agriculture in nineteenth-century Devon and west Somerset, she has worked and published on the history of agricultural science in Britain. Her current research interests are in the field of environmental history and the history of science. She now works as part of a team engaged in editing Charles Darwin's correspondence and is an associate editor of *Environment and History*.

Joyce Youings, BA PhD FRHistS was until 1987 Professor of English Social History at the University of Exeter. Her publications include *The Dissolution of the Monasteries* (1971) and *Sixteenth-Century England* (1984) and she was one of the editors of *The New Maritime History of Devon*, 2 vols, 1992 and 1994. She was for forty years one of the general editors of the Devon and Cornwall Record Society.

ACKNOWLEDGEMENTS

The editor would like to record his thanks to the University of Exeter's Centre for South-Western Historical Studies and the University of Exeter Press. The advice and support of Roger Kain, Robert Higham and Jonathan Barry of the University of Exeter both in connection with arrangements for the Centre's eighth annual symposium (which focused on Topographical Writers in South-West England), and subsequently during the production of this volume, is most gratefully acknowledged. Indeed the expert comment and advice of Dr Barry in his capacity as General Editor of the Exeter Studies in History series has been invaluable. Thanks are also due to Simon Baker and Genevieve Davey of the University of Exeter Press for their kindness and support during the preparation and production of the volume.

Special thanks are due to Tim Absalom, Brian Rogers and Andrew Hoggarth of the Department of Geographical Sciences at the University of Plymouth for preparing maps and illustrations for the volume. Similar thanks to Tony Smith and David Griffiths of the University of Plymouth for their photographic expertise. Permission to photograph the illustrations which appear as Figures 6.1–6.3 was kindly given by the City Library in Plymouth. I am also most grateful to Sir Richard Carew Pole for permission to photograph the portrait of his Tudor forebear, the topographical writer Richard Carew, which hangs in the hall of Antony House in Cornwall. Richard Carew's portrait appears on the cover of this volume. Thanks are also due for the considerable assistance in searching computerised bibliographies and thereby helping in the preparation of the listing of topographical works which was afforded by Andrew Collins of the Department of Geographical Sciences at the University of Plymouth. The editor is also grateful to Mark Cleary and Brian Chalkley of the University of Plymouth for their comments on earlier versions of the text, particularly the introductory chapter.

A large number of others have helped the editor and especial thanks are due to many staff of the university and public libraries of South-West England, staff of the Bodleian Library, the British Library and the Public Record Office. In particular, thanks to Ian Maxted of the Westcountry Studies Library in Exeter, Ann Landers the Local and Naval History Librarian in Plymouth, Angela Browne of the Courtney Library of the Royal Institution of Cornwall, Margery Rowe of the Devon Record Office, Hugh Jaques of the Dorset Record Office, Christine North of the Cornwall Record Office, and Adam Green and Tom Mayberry of Somerset Record Office.

Finally, the editor expresses his sincere thanks to the authors who have contributed chapters to this volume. I am indebted to them for their cooperation, patience and, above all, their scholarship.

Preface

The idea for this book was developed in the spring of 1993 when the University of Exeter's Centre for South-Western Historical Studies was finalising plans for its eighth annual symposium which was to be devoted to the theme of antiquarian and topographical writers in the South West. Professors Todd and Youings, and Drs Bettey, Dunning and Wilmot had already been persuaded to present papers to the meeting scheduled for November that year. Moreover, Dr Chandler had been invited to speak in October at the Royal Albert Memorial Museum in Exeter about John Leland's 'itinerary' in the West Country. An opportunity to collect in one volume the contributions of such a distinguished group of experts was simply too good to miss. It was, of course, known that a festschrift to Christopher Elrington, the former Editor of the *Victoria County History*, which was to be devoted to 'English County Histories', was already well advanced and would certainly be published within months of our Exeter symposium. Indeed, Professor Youings, Dr Bettey and Dr Dunning had contributed chapters to it on Devon, Dorset and Somerset respectively. An early question was therefore whether a book devoted to the topographers of South-West England would merely duplicate the relevant chapters in the 'Elrington festschrift'. A key consideration was that the latter volume was designed to devote no more than 16,000 words (4,000 words per chapter) to the South-West counties in a book containing well over forty chapters. In such circumstances, it was clear that the book planned as a vehicle for the Centre's symposium papers and John Chandler's Exeter lecture, while remaining relatively modest in scale and price, could nonetheless allow a much more generous treatment of this region's past topographers and antiquarians than has been attempted elsewhere. Even those authors who have contributed a chapter in both books have sought to 'ring the changes' and to introduce new ideas and extra material in the present volume. Moreover, several of the chapters contributed to this volume depart from the 'county-by-county' model (which has been so frequently used in earlier discussions of topographical and antiquarian writing) and provide thereby a broader overview of developments in this particular writing genre. Certainly, it is hoped that the book will serve not only to fill a scholarly gap, but also to draw attention to the important legacy of topographical works produced by past writers in this region. Finally, and perhaps most importantly, I hope this book will stimulate its readers to develop an interest in the historiography of topographical study in the South West and to seek out and read afresh the works focused upon the landscapes of our region which have been produced by generations of past scholars.

Mark Brayshay
Plymouth
St Bartholemew's Day
24 August 1995

1

INTRODUCTION

THE DEVELOPMENT OF TOPOGRAPHICAL WRITING IN THE SOUTH WEST

MARK BRAYSHAY

Each new generation of historians is heir to the work of those that went before and though today's scholars seek to rediscover and re-interpret the past for themselves, their endeavours are inevitably built on the foundations of an extensive legacy of previous studies. In consulting the work of earlier writers, it behoves today's researcher not only to question the evidence used by their predecessors, and to re-evaluate the conclusions based upon it, but also to know something of the life and personality of past authors and the contexts within which their studies were carried out. Complete objectivity may be a worthy modern academic goal, but even today no practitioner can escape entirely the influence of the social and cultural milieux in which he or she operates. And for earlier writers, the eradication of personal prejudices and biases was not necessarily an overriding imperative. Whether consciously or not, a writer's perception will be refracted from a particular point of view. We are, after all, creatures of our own time and place, shaped by prevailing fashions and preoccupations. Students of history must therefore seek knowledge not only of their primary sources, but also of the studies made by important previous scholars, and understand both in their proper context.

The primary purpose of this book is to identify and explore the work of key 'topographical' writers of the past who focused their attention on the landscape and society of South-West England; in particular it deals with those who wrote about the counties of Cornwall, Devon, Somerset and Dorset. In November 1993, the University of Exeter's Centre for South-Western Historical Studies devoted its eighth annual symposium to the theme of antiquarian and topographical writers in South-West England, *c*.1600–1900, and five of the six chapters presented here are adapted, and in some cases extended, essays based on the papers presented at that time. However, Chapter 2, by John Chandler, which discusses the topographical descriptions of the South West by perhaps the greatest Tudor antiquary, John Leland, was presented separately as a paper in October 1993 to a meeting of the Devon Archaeological Society held in the Royal Albert Memorial Museum in Exeter. In Chapter 3, Joyce Youings identifies and evaluates the work of the leading Tudor and early Stuart topographers who described Cornwall and Devon. In Chapters 4 and 5, Robert Dunning and Joseph Bettey contribute similar essays focused on Somerset and Dorset respectively, though both extend their chronological coverage into the modern era. Malcolm Todd's essay forms Chapter 6 and sets in a wider context the particular, and exceptionally valuable early nineteenth-century contribution of the remarkable Lysons brothers. Professor Todd shows how their work was part of a change whereby speculative studies about ancient settlement in Britain were replaced and eclipsed by archaeological investigations. Finally, in Chapter 7, Sarah Wilmot provides a detailed examination and interpretation of the topographical content of the impressively large corpus of work produced in the late eighteenth and nineteenth centuries by agricultural writers who examined the condition and potential of the farming landscapes of the counties of South-West England.

No claim is made that a comprehensive coverage is achieved here of all past texts which describe the topography of the region and its constituent counties; the intention instead is to discuss the character and content of some of the best of those available, to describe the background circumstances in which they were written, to identify their contribution and their weaknesses, and to point to ways in which they may be of use to the modern researcher. There is, of course, no substitute for consulting directly the works described by the contributors to this book. Fortunately, the majority are available in published form, but some exist only in manuscript and are thus less accessible. Overall, however, the South West can boast a splendid inheritance of four centuries of topographical studies and the best are perspicacious, incisive eye-witness accounts which can often breathe life anew into long-vanished societies, and make tangible the townscapes and landscapes of past times. In order to set the wider context for the collection of essays which follows, this introduction will examine the legacy of topographical studies of the counties of South-West England as part of the general evolution of the genre from its inception, as a well-defined form of scholarly enquiry in the Tudor period, until the final quarter of the nineteenth century.

Topographical Writing: A Developing Agenda and Spatial Focus

When he introduced his splendid and much-admired topographical *Survey of Cornwall*, first published in 1602, Richard Carew of Antony (1555–1620) informed his readers that, 'this treatise plotteth downe Cornwall, as it now standeth, for the particulars, and will continue, for the generall. Mine *Eulogies* proceede no lesse from the sinceritie of a witnesse, then the affection of a friend.'[1] In making this statement, Carew was clearly aligning his work with the developing enthusiasm for eye-witness topographical descriptions, customarily combined with outline historical accounts, written by educated men of the Tudor and early Stuart period who wished both to celebrate the land they loved and admired, and to define its essential character.[2]

Although topographical descriptions (focused on small tracts of landscape) and 'chorographical' descriptions (focused on well-defined regions) were not entirely new in Tudor England, from the early sixteenth century there was certainly a remarkable growth of interest in this kind of enquiry: 'Almost suddenly the notion was born that all around . . . was this unknown country . . . and maps, descriptions, and histories, poured forth for the next 300 years'.[3] The growing appetite for such work was related in part to the wider European cultural renaissance, but other factors also played a role. The crystallisation of the nation and a growing sense of nationhood, the spread of literacy and education, and the recognition of a quickening in the pace of change: all undoubtedly exerted an important influence.[4]

Although John Leland and William Camden each managed, in their respective surveys made during the sixteenth century, to encompass the country as a whole, given the increasingly accepted requirement that topographical and chorographical studies should be based on first-hand knowledge as well as antiquarian research, the scale of such an undertaking increasingly lay beyond both the intellectual and physical capacity of a single author. Indeed, the burden of the huge task he had set himself ultimately may have cost John Leland his sanity.[5] For a variety of reasons, John Norden's ambitious *Speculum Britanniae* project, which was intended as a 'Chorographicall description of the severall Shires and Islands' of the realm, was never fully realised, although (to the great good fortune of South-West England) his *Speculi Britanniae Pars. A Topographicall and Historical Description of Cornwall* was completed, together with its set of sumptuous maps.[6] Gradually, however, most Tudor topographers narrowed the scale of their work to focus upon much smaller areas. In so doing they not only rendered their project more manageable, but also thereby tacitly recognised the wisdom of William Lambarde who wrote in his *Perambulation of Kent*, published in 1576: 'The inwards of each place may best be known by such as reside there . . . whereby good particularities will come to discovery everywhere . . . to amplify and enlarge the whole'.[7] Nevertheless the predominantly 'county focus' of Tudor and Stuart topographical studies requires some further explanation. Long in existence to define the units of English local government, county boundaries tend to identify appropriate administrative rather than topographical divisions. However, by the sixteenth century, the county was increasingly playing a crucial government role whereby the local gentry were appointed to key offices such as that of deputy lieutenant or justice of the peace with responsibility, on the Crown's behalf, for local justice, administration and defence.[8] Inevitably interested in the antiquity and pedigree of their own family and estates, and their association with the locality, the more scholarly gentry, some of whom held Crown appointments in their county, frequently took a lively interest in local history. The more literary-minded recorded their observations and took the county as the convenient focus of their surveys. In fact, by 1800, all but a handful of English counties had been 'described'.[9] The character and content of such studies naturally evolved and changed over the decades but, by the 1870s, interest had somewhat waned and there was a hiatus until the launch, in 1899, of the *Victoria County History* which aimed to produce a historical encyclopaedia for every English county. Almost a century later, work on this great national project to write the local history of the country is still in progress.[10]

The Evolution of Tudor Topographical Studies

The origins of Tudor topographical surveys may be traced to the late fifteenth-century observations recorded by William of Worcester, or William Botoner, who was born in Bristol in 1415. He studied at Oxford before becoming a retainer to Sir John Fastolf of Caistor Castle in Norfolk in 1432. In 1478 William travelled from Norwich to St Michael's Mount in Cornwall and his manuscript account of that journey survives in the library of Corpus Christi College, Cambridge. The highlight is a street-by-street description of his home town of Bristol. Though somewhat skeletal and unsystematic in his approach, William was a great fact-gatherer who noted, for example, the distances between towns, the course of rivers, the number of bridges, and the locations of castles, churches, havens and ports. In places his text has an immediate quality which conjures up a vivid image of a fifteenth-century journey in progress: 'Sunday the 20 September [1478], I rode from Bodmin to the town of Lostwithiel and came to the town of . . . Fowey. Talked and abode the night with Robert Bracey.'[11] Elsewhere it is clear that William sought detailed local information such as that, for example, on the dimensions and character of Lundy Island: 'The Island of Lundy, on the Severn Sea board, lies in the south part, twenty one miles in the sea . . . and contains in length three miles, and in breadth three miles. It is not peopled.'[12] It appears that William of Worcester may have been deliberately collecting material for a 'Description of Britain', but no such work was produced and he died in 1490. In fact his manuscript lay neglected until the eighteenth century when an edition was published by Nasmith.[13] However, William's work set the tone for the surveys made by many of the Tudor topographers who came after. In particular, the majority of later studies are arranged as if the writer was following a real or imaginary itinerary. By describing their peregrinations, writers inevitably mention a great many places by name, but sometimes, it must be said, at the expense of detailed commentary or a general overview.

Notwithstanding the role of William of Worcester, John Leland (1503–52) is generally regarded as the earliest antiquary and topographer.[14] As the chapter by John Chandler in this book clearly shows, his contribution and influence was considerable. Leland's 'Itinerary' demonstrated that to be acceptable and authoritative, descriptions had to be not only detailed and comprehensive, but also based as much as possible upon original observations in the field. His own words reveal a fanatical attachment to his native country:

> I was totally enflammid with a love to see thoroughly al those partes of this your opulente and ample reaulme, that I had redde of yn the aforesaid writers: yn so muche that al my other occupations intermittid I have travelid yn yowr dominions booth by the se costes and the midle partes, sparing nother labor nor costes, by the space of these vi. yeres paste, that there is almost nother cape, nor bay, haven, creke or peere, river or confluence of rivers, breches, waschis, lakes, meres, fenny waters, montaynes, valleis, mores, hethes, forestes, wooddes, cities, burges, castelles, principale manor placis, monasteries, and colleges, but I have seene them; and notid yn so doing a hole worlde of thinges very memorable . . . I truste that this yowr reaulme shaul so welle be knowen, ons payntid with his native coloures, that the renoume ther shaul gyve place to the glory of no other region.[15]

In the South West of England, Leland's 'Itinerary' and his other works reveal a landcape still largely medieval in character: the visual manifestation of a feudal and military past. His approach, based on a blend of first-hand recording, questioning of local people, quarrying the written sources, and research in the archives of the greater gentry houses, set a standard to which many later Tudor antiquarian and topographical writers aspired.

Following Leland, William Harrison attempted a new topographical *Description of Britain* and *Description of England*, probably written in the late 1560s or early 1570s, but published as an introduction to Ralph Holinshed's *Chronicles* in 1577.[16] Harrison's work was not based on eyewitness observations; indeed he sheepishly acknowledged that he had not 'by mine own travel and eyesight viewed such things as I do here entreat of'.[17] Foreshadowing a practice that became increasingly widespread, Harrison in fact relied heavily upon Leland's text, though he noted that when he read them, the manuscripts were already in a deteriorated condition and some had been lost altogether. But there is much useful new topographical information on the West Country in Harrison's *Description* and his discursive style stands as an interesting contrast to that of the colder, clear-headed observations of William Camden (1551–1623).

Camden's extraordinary work, *Britannia, sive Florentissimorum Regnorum Angliae, Scotiae, Hiberniae, et Insularum adjacentium ex intima antiquitate Chorographica Descriptio*, was first published in 1586 with a dedication to Lord Burghley.[18] After leaving Oxford in 1571, Camden travelled around the country collecting information. In 1575 he began his career as master, eventually headmaster, of Westminster School, where he remained until 1597, an occupation which enabled him to travel and work on the *Britannia* during vacations.[19] In addition to his fieldwork, he is known to have carried out considerable documentary research and to have incorporated information contributed by local correspondents. But like Harrison, William Camden drew heavily on the manuscripts of John Leland. Indeed, blatant plagiarism was by then well-established in antiquarian and topographical writing. Mistakes and misunderstandings could thus enjoy remarkable longevity. But as a result of further indefatigable research, Camden was able to extend the scope and length of his *Britannia* through five editions in Latin, crowned by the folio version published in 1607 which was further enhanced by the inclusion of copies of the maps of Christopher Saxton and John Norden.[20] The first English translation appeared in 1610 and is regarded as marking an intellectual and cultural turning point. It has been argued that the English by then had:

> moved out of that Latin-speaking fraternity of learning which, up to the time of Elizabeth, had carried on the tradition of the scholars' *lingua franca*, and are [by Jacobean times] in the new, self-confident nation state in which, with the increase in literacy, an interest in local history was no longer confined to the learned professions, but was as likely to be found in the merchants as in the country squire.[21]

Camden's organisational paradigm was based on celtic tribal areas and how these fitted within, or straddled, the English shires. Although emphasising the ancient origins of the landscapes he describes, the *Britannia* is also a rich source of topographical detail. Perhaps the best-known version of the *Britannia* is that edited by Bishop Edmund Gibson in 1695 and illustrated by the maps of Robert Morden, generally referred to as *Gibson's Camden*.[22]

Often regarded as the pinnacle of Tudor and early Stuart topographical writing, the *Britannia* marks only the beginning. Camden recognised this when he wrote that, 'I have broken the ice; and I have gained my end if I set others to work'.[23] An element in the process of 'setting others to work' was Camden's involvement in the Society of Antiquaries which was established in the 1580s and met in the London home of Sir Robert Cotton where members shared supper and then heard and discussed a scholarly paper.[24] Amongst the other members were John Stow (best known for his *Survey of London*, published in 1598), Sir John Doddridge and (as Joyce Youings notes in her chapter) Richard Carew of Antony in Cornwall. Of all the work of the Tudor and Stuart topographers in the South West, Carew's *Survey of Cornwall* is easily the best. Regarded by some as a minor classic in the English language, it ranks amongst the finest of its type in Britain as a whole.[25] It is, moreover, one of the most accessible of the early works, as it was (unusually) published in Carew's own lifetime (Figure 1.1). Although in 1606 Carew was planning a second, revised edition to be accompanied, he hoped, by a map by John Norden, sadly, due to a lack of patronage and financial support, his ambition was not realised.

Joyce Youings notes that the 'companion-piece' to the *Survey of Cornwall* was the 'Synopsis Corographical of the county of Devon' by Carew's contemporary, John Hooker (1525–1601), but this has never been fully published.[26] In its manuscript form, Hooker's 'Synopsis' lacks the literary quality of Carew's *Survey*, but it is nonetheless equally rich in social and topographical information. Hooker was a particularly sharp-eyed observer of the Tudor gentry and his comments shed much light on county society and local administration during the reign of Elizabeth.

For a topographical description of late Tudor and early Stuart Somerset and Dorset, attention turns to the work of Thomas Gerard (1582–1634) who prepared a *Particular Description of the County of Somerset* (which actually covers only the southern half of the county

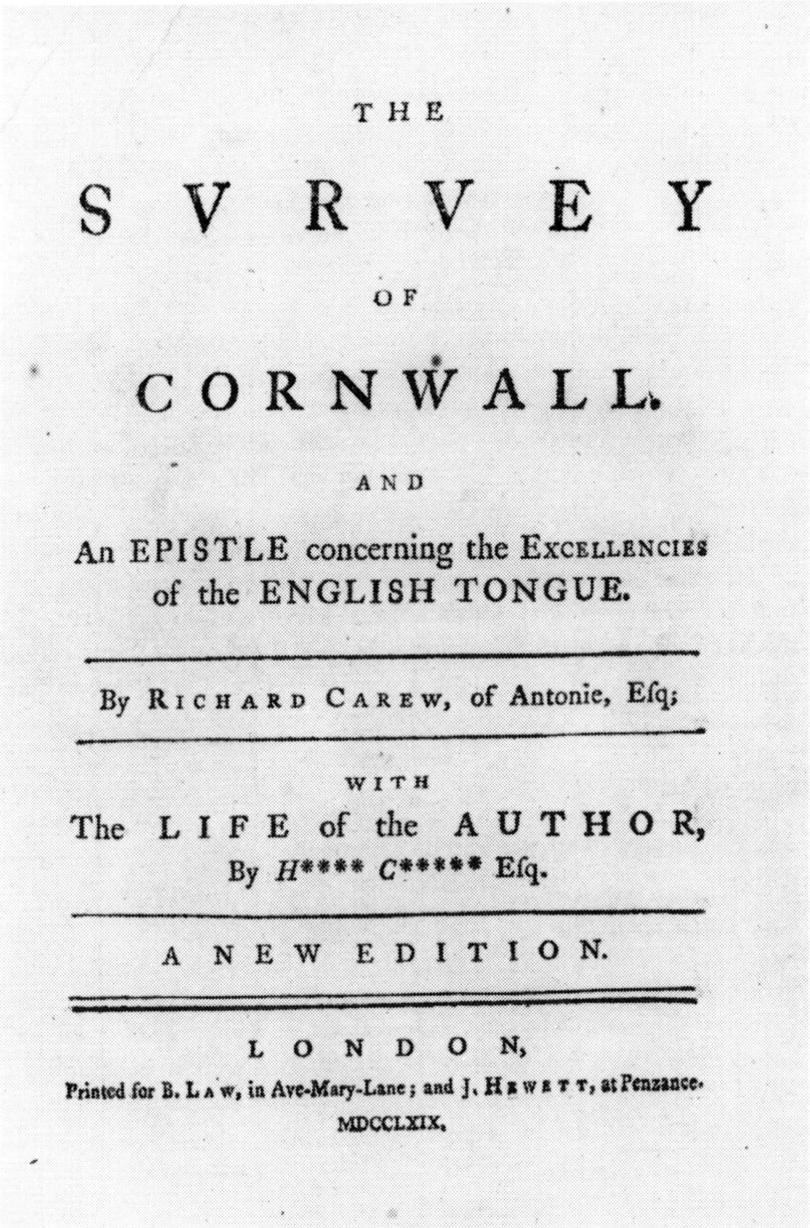

THE

SVRVEY

OF

CORNWALL.

AND

An EPISTLE concerning the Excellencies of the ENGLISH TONGUE.

By Richard Carew, of Antonie, Esq;

WITH

The LIFE of the AUTHOR,

By *H**** C****** Esq.

A NEW EDITION.

LONDON,

Printed for B. Law, in Ave-Mary-Lane; and J. Hewett, at Penzance.

MDCCLXIX.

Figure 1.1 Title Page of 1769 edition of Richard Carew's *Survey of Cornwall*, first published in 1602.

and bears the date 1633) and a *General Description of Dorset*, probably written in the 1620s.[27] Gerard's survey of Somerset is described below by Robert Dunning, while Joseph Bettey discusses Gerard's work on Dorset. Bettey's chapter also draws attention to the work of Thomas Fuller, vicar of Broadwindsor, Dorset, who published his *History of the Worthies of England* in 1662, but Fuller provides very little material on the topography of Dorset.[28] The South West is fortunate that two further Stuart authors produced historical and topographical descriptions of Devon, though both relied heavily on the manuscript description written by Hooker. Thomas Westcote (?1567–*c*.1640) completed a manuscript 'Survey of Devon' in about 1630 (which was edited and published much later), and Tristram Risdon (*c*.1580–*c*.1640) produced a 'Chorographical Description or Survey of the county of Devon' which appears to have circulated in several manuscript versions (bearing several different early seventeenth-century dates) before its eventual publication in a somewhat altered form in the eighteenth century.[29] These surveys, and what survives of the manuscripts of Sir William Pole (1561–1635) following the losses sustained during the Civil War (published in 1791), are further described and assessed in the chapter by Joyce Youings.[30]

Natural History and the New Antiquarianism

Notwithstanding William Dugdale's magnificent *Antiquities of Warwickshire Illustrated*, which he closely modelled on Camden's approach and published in 1656, by the mid-seventeenth century there had been a marked shift of focus towards descriptions of the natural history of English counties where the emphasis was on scientific observation and experiment, and on taxonomical and deductive studies.[31] Scholarly endeavours in England were increasingly informed and shaped by the humanist tradition of the Italian and wider European Renaissance which advocated a new secular scheme of organisation of scientific knowledge.[32] The Elizabethan Society of Antiquaries had been disbanded during the reign of James I, but by 1645 meetings of the new humanist scholars began to take place in London and Oxford. However, they flowered fully under the auspices of the Royal Society, founded in 1662.[33] A leading practitioner of the new approach, and a key member of the Royal Society, was John Aubrey (1626–97). Although rather better known for his field archaeology, Aubrey produced the first natural history of an English county, namely Wiltshire.[34] It has been argued that Aubrey's remarkable talent was never fully realised, but his ideas undeniably defined a new direction in topographical writing. He favoured the division of the landscape for the purposes of study into 'natural tracts' based on physical geography, soils and land capability. Further evidence ought, he felt, to be gathered by means of a questionnaire. His own manuscript list of nineteen 'queries' survives in the Bodleian Library, and his approach in systematically gathering information from local 'informants' was taken up by an number of later workers, including John Ogilby.[35] Aubrey also recognised the value of thematic maps and proposed that a soil map and a land-use map of Wiltshire were needed. His example was followed by Robert Plot (1640–96) who also used the questionnaire approach by then being advocated by the Georgical Committee of the Royal Society. Plot's work on the natural history of Oxfordshire and, later, of Staffordshire emphatically rejected the old-style antiquarian preoccupations of men like Hooker, Risdon and Gerard.[36] Indeed, Plot clearly spelled out the new direction which he was taking:

> I intend not to meddle with the *pedigrees* or *descents* either of *families* or *lands* . . . It being indeed my Designe . . . to omit, as much as may be, both *persons* and *actions*, and chiefly apply my self to *things*; and amongst those too, only of such as are very remote from the present Age.[37]

Thus, by the later seventeenth century, the term antiquarianism was itself being redefined. In Elizabeth's reign an antiquary was simply a student of history, usually of the remote past. There was, however, a distinct emphasis on genealogical and heraldic descents. Although such interests continued, studies now tended to focus on the antiquity of the landscape itself and the ancient monuments which survived within it. In addition county surveys assumed an overtly practical aim, namely to describe and assess natural resources with a view to their exploitation by agriculture, industries and trade. This new angle found expression in the revisions to William Camden's *Britannia* which were made under the editorship of Edmund Gibson and, as noted earlier, published in 1695.

Gibson assembled a group of experts who each contributed to the new additions. For example, Dr Edward Lhwyd, Robert Plot's successor as keeper at the Ashmolean Museum in Oxford, prepared the natural history material for the Cornwall sections of *Gibson's Camden*.

In South-West England, the earliest work that accords most closely with the ideas of the natural historians and the 'new antiquarianism' was that of John Beaumont (d. 1731) who, as Robert Dunning reports below, proposed a new study in 1685 of the 'History of Nature and Art' in Somerset, but sadly the project was never realised.[38] Other writers representing both kinds of antiquarianism produced studies of South-Western counties which include traditional eye-witness topographical descriptions. In 1737 Thomas Tonkin proposed a history of Cornwall in three volumes, and although this was never published, it survives in manuscript form in the British Library.[39] Tonkin also prepared an account in alphabetical order of all the Cornish parishes (but completed only those from A to O); the manuscript eventually passed to William Borlase.[40] Another early eighteenth-century attempt to produce a complete set of parochial histories for Cornwall was made by William Hals. Again, part of his work survives only as a manuscript in the British Library, but descriptions of seventy-two parishes were published by Andrew Brice of Truro in 1750 as the *Compleat History of Cornwall*.[41]

Best known amongst Cornwall's eighteenth-century scholars is William Borlase (1696–1772), who was born in the parish of St Just in Penwith and, after several years at Exeter College, Oxford, became vicar of Ludgvan, near Penzance. His *Natural History of Cornwall*, published in 1758, is his most valuable work of topographical description and by devoting more than half of the text to geology and mineralogy, Borlase showed that his interests were firmly rooted in the tradition of natural historical studies developed in later Stuart times.[42] Nevertheless, the scope of Borlase's *Natural History* is epic. In addition to the lengthy account of the geology and natural resources, it encompasses the county's climate, river systems, coasts, flora and fauna. Moreover, it describes the population, the economy, Cornish culture and language. In 1766, Oxford ultimately acknowledged the considerable contribution which Borlase had made to scholarship by awarding him an honorary doctorate.

Borlase's other great contribution was to the study of the ancient Cornish landscape and in 1754 he had published his *Observations on the Antiquities of the County of Cornwall*, illustrated with his own detailed drawings of archaeological monuments and artefacts, and plans of key sites.[43] A second edition, with a slightly altered title, followed in 1769 and the work is often regarded as establishing the foundation of archaeological research in the county (Figure 1.2). In 1756 Borlase produced a natural history of the Isles of Scilly entitled *Observations on the Ancient and Present State of the Islands of Scilly* (based on surveys made during a visit to the islands in 1752). By then he was clearly at the height of his intellectual powers and his work was commended in the *Literary Magazine* as 'one of the most pleasing and elegant pieces of local enquiry that our country has produced'.[44]

For Somerset, the county history by John Strachey (d. 1743) which he intended, but failed, to publish survives in manuscript form in the Somerset Record Office.[45] John Collinson's *History and Antiquities of Somerset* is more accessible, however, as it was published in three volumes in 1791–2.[46] As Robert Dunning notes, Collinson had been curate at Marlborough before moving to Cirencester where he remained until 1788. From then until his death in the 1790s, he was vicar of Long Ashton in Somerset.

The first attempt to write a full history of the county of Dorset was made by Thomas Cox in his *Compleat History of Dorsetshire*, published in 1730 as part of his six-volume *Magna Britannia* (1720–31).[47] This was followed by the work of John Hutchins (1698–1773) on the *History and Antiquities of the County of Dorset* which was published in two volumes a year after his death.[48] Two further, enlarged editions were subsequently published.

The Revd Richard Polwhele (1760–1838) was born in Cornwall and educated at Truro Grammar School before going up to Christ Church, Oxford. But he left before taking his degree, and returned to Cornwall where he was ordained, married, and made curate of Lamorran. Shortly after, however, he moved to the curacy of Kenton, near Exeter, where he remained for ten years. During that period he began collecting material for a history of Devon. He published a list of 'Queries' for his proposed 'History' in the *Gentleman's Magazine* in 1791, thereby employing the questionnaire approach first advocated more than a

ANTIQUITIES,

HISTORICAL and MONUMENTAL,

OF THE

COUNTY of CORNWALL.

CONSISTING OF

SEVERAL ESSAYS

ON

The FIRST INHABITANTS, DRUID-SUPERSTITION, CUSTOMS,
And REMAINS of the moſt Remote ANTIQUITY
In BRITAIN, and the BRITISH ISLES,
Exemplified and proved by MONUMENTS now extant in CORNWALL
and the SCILLY ISLANDS,
With a VOCABULARY of the CORNU-BRITISH LANGUAGE.

By WILLIAM BORLASE, LL.D. F.R.S. Rector of LUDGVAN,
CORNWALL.

" Miratur, facileſque oculos fert omnia circùm
" Æneas, capiturque locis, et ſingula lætus
" Exquiritque auditque virûm Monumenta priorum." VIRG.

The SECOND EDITION, reviſed, with ſeveral Additions, by the Author; to which is added a Map
of CORNWALL, and Two new Plates.

L O N D O N,

Printed by *W. Bowyer* and *J. Nichols*:
For S. BAKER and G. LEIGH, in York Street; T. PAYNE, at the Mews Gate, St. Martin's;
and BENJAMIN WHITE, at Horace's Head, in Fleet Street.
MDCCLXIX.

Figure 1.2 Title Page of William Borlase's *Antiquities, Historical and Monumental, of the County of Cornwall*, 1769.

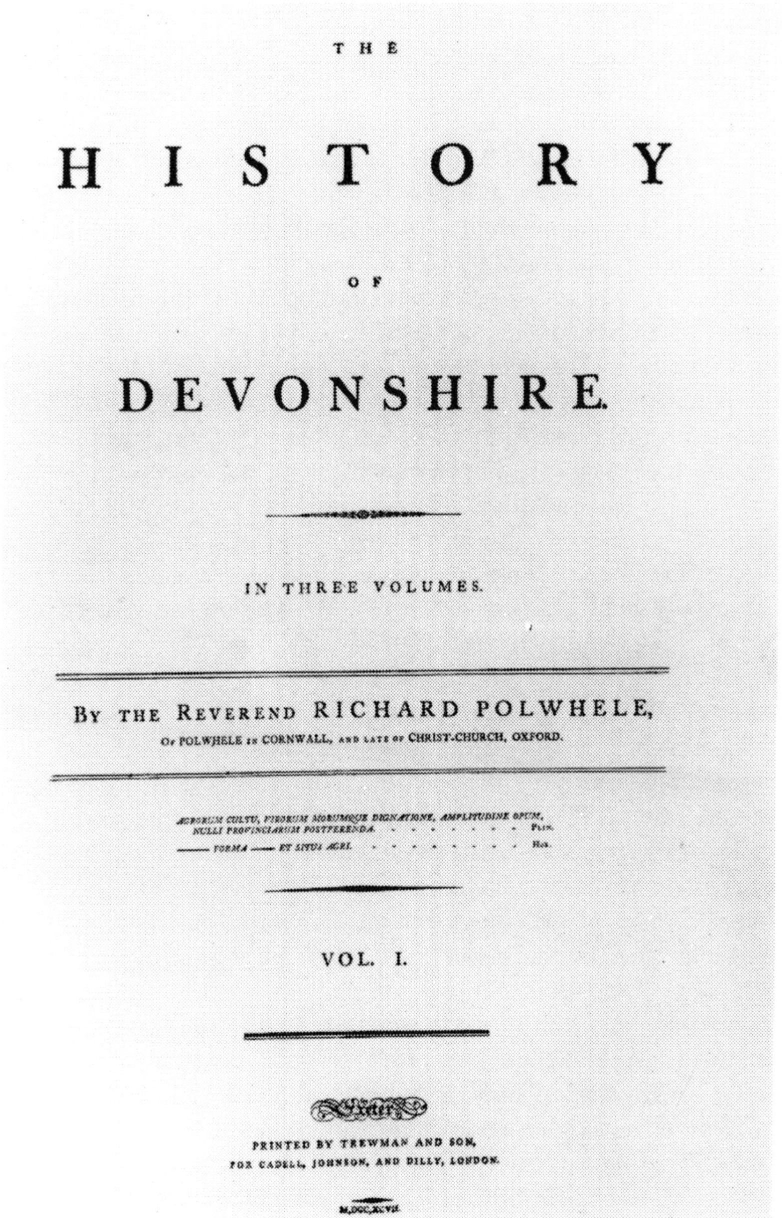

THE

HISTORY

OF

DEVONSHIRE.

IN THREE VOLUMES.

BY THE REVEREND RICHARD POLWHELE,

OF POLWHELE IN CORNWALL, AND LATE OF CHRIST-CHURCH, OXFORD.

AGRORUM CULTU, VIRORUM MORUMQUE DIGNATIONE, AMPLITUDINE OPUM,
NULLI PROVINCIARUM POSTFERENDA. - - - - - - - - PLIN.

—— *FORMA* —— *ET SITUS AGRI.* - - - - - - - - - HOR.

VOL. I.

PRINTED BY TREWMAN AND SON,
FOR CADELL, JOHNSON, AND DILLY, LONDON.

M,DCC,XCVII.

Figure 1.3 Title Page of Richard Polwhele's *History of Devonshire*, 1797.

century earlier by John Aubrey and subsequently by the Georgical Committee of the Royal Society.[49] His three-volume *History of Devonshire* appeared between 1793 and 1797 (Figure 1.3), followed by a revised edition in 1809.[50] The first part of Polwhele's *Devonshire* comprises a remarkably detailed and fluently written 'sketch of the natural history' of the county which contains (amongst many other delights) a detailed account of Devon's weather based on observations made by the vicar of Cornwood who had 'accurate instruments for that purpose'. The remainder of the first volume comprises a history of Devon focusing especially on the role of the county in the history of the nation as a whole. In the second and third volumes Polwhele offers a lengthy series of parish descriptions, grouped by deanery rather than by hundred. His style is elegant and engaging and he repeatedly shows a talent for encapsulating an idea or judgement in a well-turned phrase. Thus, he sums up Devon's scenic qualities by saying, 'even in winter, this County seems to possess the more agreeable charms of landscape—such as no other part of the island presumes to emulate'. But, by 1794, Polwhele had returned to Cornwall, to be curate of Manaccan, where he remained until 1806 when he secured the curacy of Kenwyn, near Truro. His final appointment was to Newlyn East in 1821. Amidst his many other activities, not least as father of more than a dozen offspring, Polwhele turned his attention to a history of his native Cornwall. This was eventually published in three volumes in 1803, but then republished as an enlarged and improved work in seven volumes in 1816.[51] In contrast to his work on Devon, Polwhele's Cornwall does not comprise parish-by-parish descriptions, but is instead arranged thematically. He provides, for example, detailed discussions of the local economy, the population (drawing on the 1801 census data), and health and housing in Cornwall.

Polwhele's magnifient studies of Devon and Cornwall assemble information which is not easily obtainable elsewhere and, as eye-witness accounts of the late Georgian landscapes which he knew so intimately, they scarcely have an equal. For years Richard Polwhele was a somewhat neglected topographical writer, but his work is now more widely appreciated for the considerable merit that it undoubtedly possesses.

Daniel and Samuel Lysons' *Magna Britannia: 3 Cornwall* was published in 1814 and includes both a general historical and

MAGNA BRITANNIA;

BEING

A CONCISE TOPOGRAPHICAL ACCOUNT

OF

THE SEVERAL COUNTIES

OF

GREAT BRITAIN.

By the Rev. DANIEL LYSONS, A.M. F.R.S. F.A. and L.S.

RECTOR OF RODMARTON IN GLOUCESTERSHIRE;

And SAMUEL LYSONS, Esq. F.R.S. and F.A.S.

LATE KEEPER OF HIS MAJESTY'S RECORDS IN THE TOWER OF LONDON.

VOLUME THE SIXTH,

CONTAINING

DEVONSHIRE.

LONDON:

PRINTED FOR THOMAS CADELL, IN THE STRAND.

1822.

Figure 1.4 Title Page of Daniel and Samuel Lysons' *Magna Britannia, Volume 6: Devonshire*, 1822.

topographical survey and detailed parochial histories of the county.[52] Their work on Devon was to be the last in the series. *Magna Britannia: 6 Devon* was published by Daniel Lysons in two volumes in 1822 after the death of his brother, Samuel (Figure 1.4). Daniel, rector of Rodmarton in Gloucestershire, was deeply affected by the loss of his lawyer-archivist brother, who had been keeper of the records in the Tower of London: 'The publication of the present volume has been thus long delayed principally in consequence of the melancholy loss sustained in 1819 by the death of my brother. It was a considerable time before I could feel equal to resume the work.'[53] In format, the Devon study matches that of Cornwall and both works describe the history, civil administration, ecclesiastical organisation, parishes, monasteries, colleges, ancient hospitals, boroughs, markets, and fairs of the county. They describe the population, the nobility and the gentry; they enumerate the forests and deer parks and the country estates. Physical geography, natural history, resources, trade, and manufacturing are also discussed. Parochial histories are included and, as the chapter by Malcolm Todd indicates, the Lysons brothers provide detailed descriptions of archaeological antiquities. Samuel's earlier work on archaeological sites in Gloucestershire established a blueprint for the archaeological sections of their *Magna Britannia*. The latter is adorned with sumptuous illustrations provided by a variety of artists, including Samuel Lysons himself.

Other notable nineteenth-century surveys made by natives or residents of the South-Western counties which contain topographical description include Charles Sandoe Gilbert's two-volume *Historical Survey of the County of Cornwall*, published between 1817 and 1820.[54] As a travelling purveyor of pharmaceutical products, Gilbert differed from most other nineteenth-century historians and topographers who tended to be Anglican clergymen. Their calling appears to have afforded them sufficient spare time to pursue their antiquarian and topographical interests, whereas the unfortunate Gilbert's did not. Indeed, his neglect of his business ultimately resulted in bankruptcy in 1825.[55] By contrast, the Revd Thomas Moore, whose *History and topography of the county of Devon* was published in 1831, was more typical of that significant group of amateur historians drawn from amongst the local clergy who described the history, the genealogy of notable families, and the contemporary topography of the counties of the South West in the nineteenth century.[56]

Davies Gilbert, or Giddy, published his four-volume *Parochial History of Cornwall* in 1838, a work based largely upon the manuscript histories of Hals and Tonkin.[57] Indeed, most of the nineteenth-century topographers demonstrate an intimate knowledge of the manuscripts and publications left behind by earlier scholars. But Gilbert's health problems obliged him to rely on the assistance of collaborators, who were often not sufficiently assiduous in their efforts, with the result that the value of the work was undermined. Joseph Polsue's work followed what by then had become a well-established pattern of alphabetical parish histories and was published as the *Complete Parochial History of Cornwall* in four volumes between 1867 and 1872.[58] This work is sometimes known as *Lake's Parochial History*, after one of its several publishers.

Some Victorian topographers focused detailed attention on particular districts rather than making an attempt to encompass an entire county. One of the most notable was the masterly study of Dartmoor produced in 1848 by the Revd Samuel Rowe. His *Perambulation of the Ancient Royal Forest of Dartmoor* went through several editions and has recently been reprinted again in facsimile form.[59] However, the labour required to gather sufficient information in order to write a topographical description, even when only a relatively small area was tackled, meant that few of the many projected works were actually completed. Occasionally the research papers have survived of those who tried, but ultimately failed to complete a new study. A good example is the collection of F. W. L. Stockdale who seems, in fact, to have been somewhat ambitiously planning yet another new gazetteer of all Devon's parishes and sought information from the county's clergymen. Many of the replies which Stockdale received remain in the files of the Devon and Exeter Institution and some contain valuable local topographical detail which does not exist in any other source. A full-scale analysis of their content could well prove a rich seam for future research.[60] By the mid-Victorian period, however, railway travel had already prompted an increased demand for the kind of traveller's guide book that would be easily recognisable today. Focusing not only on counties, but also on individual towns and districts, this new type of literature can often offer most useful topographical observations, though only a few rank as works of serious scholarship. However, among the best are John Murray's *Handbooks for Travellers* which began with Devon in 1850 and by 1905 had covered all the English and Welsh counties. Together Murray's *Handbooks* represent a monumental corpus of information about the landscape of nineteenth-century England and Wales.

Diarists and Observant Travellers

Although scarcely either systematic and scientific in their approach, or comprehensive in their spatial coverage, topographical descriptions made by observant travellers who toured in the counties of South-West England can provide important insights and considerable detail about the landscape of Cornwall, Devon, Somerset and Dorset. The number of useful travel diaries increases markedly in the second half of the seventeenth century. Indeed, as recent work has demonstrated, there are undoubtedly a great many more hitherto unknown records of this kind that await a full evaluation of their topographical content.[61] But as a source of genuinely objective evidence about the region's landscapes and people, the reports of travellers, perhaps above all others, need to be treated with considerable caution. On the other hand, the eye of an outsider, unclouded by the sentiment of native attachment, can sometimes produce the shrewdest observations.

A full evaluation of the topographical descriptions of all the known early tourists who visited the South West lies beyond the scope of this book, but amongst the earliest was Lieutenant Hammond who began a seven-week tour of the western counties in the company of two friends in 1635. Hammond clearly held somewhat extreme pre-conceived notions regarding the landscapes and people of Cornwall and these appear ultimately to have dissuaded him from visiting the county:

> I (after bidding farewell) then tooke Horse, and speeded on my Journey, but had no desire [to go] over Tamer, to the horned-nock-hole Land's end, nor her horned wayes to the rough, hard-bred, and brawny strong limb'd wrastling Inhabitants thereof. Nor to the north Rivers of Tow and Towridge. Away therefore I troop'd for Taunton . . .[62]

In 1669 the Grand Duke of Tuscany, Cosmo III, visited the South West and a manuscript account of the journey written by Count Lorenzo Magalotti survives in the Laurentian Library in Florence and was translated and published in 1821. By recording the impressions of a visiting Florentine courtier, this diary has a special fascination. It is, however, hardly a surprise that Magalotti devotes a level of attention comparable with that of a professional spy to local defence installations. In the case of Plymouth, he reports that:

> On the sea-side, towards the east, and near the coast, is a small isolated mountain, called St Michael's [Drake's Island], capable of defending, to a certain extent, the first entrance to the port. There are other fortifications at the mouths of the rivers; on the Tamar, is an ancient castle, called St Francis; on the highest part of a small island, and on the Plym, an entrenchment of earth, well supplied with artillery; a similar one defends the mouth of the dock, towards the city; and others are disposed on a rock which protects, in front, the whole length of the bay.[63]

Undoubtedly the best known of the observant travellers of the later Stuart period is Celia Fiennes, whose first West-Country journey took her to Dorset and Somerset between 1682 and 1685.[64] Her eye-witness accounts have a certain freshness and can be highly informative, but her judgemental tone can also pall. The mix of all three qualities in her work may be discerned, for example, in her description of Somerset cider orchards:

> In most parts of Sommer-setshire it is very fruitfull for orchards, plenty of apples and peares, but they are not curious in the planting the best sort of fruite, which is a great pitty; being so soone produced, and such quantetyes, they are likewise as careless when they make cider, they press all sorts of apples together, else they might have as good sider as in any other parts, even as good as Herriffordshire; they make great quantetyes of cider their presses are verye large, so as I have seen a Cheese, as they call them, which yeilded 2 hoddsheads . . .[65]

Celia's 'Great Journey to Newcastle and to Cornwall' was undertaken in 1698 and she entered the West Country via Bristol, travelling through

Somerset and Devon towards Cornwall, and eventually reaching Land's End. Her return itinerary to her home at Newton Toney in Hampshire brought her back via Dorset. As a member of a lesser gentry family, Celia sought accommodation whenever she could in the homes of her relatives, but was nonetheless sometimes obliged to take lodgings in a local hostelry. When such facilities failed to earn her approval, she often condemned the entire town. Such was the fate of Ashburton, dismissed as 'a poor little town', where 'bad was the best inn'.[66] The pejorative character of many of Fiennes's comments may in part be explained by her own words:

> As this was never designed, soe not likely to fall into the hands of any but my near relations, there needs not much to be said to excuse or recommend it . . . as most I converse with, knows both the freedom and easyness I speak and write, as well as my deffect in all, so they will not expect exactness or politeness in this book.[67]

Though possibly somewhat disingenuous, Celia's statement nonetheless leaves no doubt that she was aware that her bluntness might be deemed a defect.

Daniel Defoe appears to have commenced his collection of information, which was ultimately woven into a narrative description of Great Britain, as early as 1683 when he worked as a travelling hosiery factor. He probably continued building the collection whenever occasion arose thereafter. He fought on the side of the Duke of Monmouth at Sedgemoor in 1685 and three years later rode with the retinue of William III on the procession towards London. His highly chequered career presented many opportunities to see the landscape and Defoe regularly travelled around the country.[68] Thus his well-known and extremely valuable *Tour thro' the whole Island of Great Britain, Divided into Circuits or Journies*, which was published in three volumes, annually from 1724 until 1726, embodies the fruits of forty years of experience.[69] But this strength is also a weakness since it is difficult to determine when precisely he actually observed the phenomena about which he writes. Moreover, Defoe's hyperbole is occasionally extreme. Of Devon he wrote: '. . . one entire county . . . so full of great towns, and those towns so full of people, and those people so universally employed in trade and manufactures, that not only it cannot be equalled in England, but perhaps not in Europe'.[70] Rather more balanced, and with a somewhat different emphasis, are the accounts of extensive travels in England made by the Irish-born bishop, Dr Richard Pococke, who visited the South West in 1750.[71] His eye for detail and his readable style considerably enhance the quality of his somewhat undervalued descriptions. Thus, for example, he reports that:

> In Cornwall and Devon shire they have few wheel carriages by reason of the steep hills, but everything is carried either on hooks on each side of the horses, which are long or short according to the nature of the burthen; they have drags for drawing up the sides of steep fields, and what wheel carriages they have are drawn by oxen and horses which they use for ploughing.[72]

Pococke often reveals a genuine appreciation for the grain of the landscape, providing clear evidence that his accounts were based on genuine first-hand observation. Thus, he described crossing the river Taw and 'having crossed it, I found another sort of face of the country, the red soil with the red sandstone, and all the country full of rising grounds, and small hills beautifully improved'.[73] Astute observations may also be found in the descriptions of the Revd Stebbing Shaw in his Tour in the West of England in 1788. For instance, he sheds light on why so few tourists describe Saltram when Mount Edgcumbe rarely escapes the observer's attention:

> we approached the vicinity of Plymouth, in which are several good seats; particularly one at Saltram, belonging to Lord Boringdon (John Parker), whose situation and hanging woods by the side of this arm of the sea might be deemed worthy [of] much attention, was there not so great a rival just opposite [Mount Edgcumbe].[74]

The manuscript travel journals of the Revd John Swete, 1789–1804, with their large collection of beautiful watercolour illustrations, form another extremely important source of topographical description for both Cornwall and Devon.[75] Amongst a range of other themes, Swete's comments on industry and agriculture are particularly valuable as he was making his observations at a time of fundamental and far-reaching change.[76]

William George Maton's *Observations relative chiefly to the Natural History, Picturesque Scenery and Antiquities of the western Counties of England, Made in the years 1794 and 1796* represent the published version of part of a very extensive collection of manuscript journals surviving in the British Library.[77] Dr Maton's approach was unmistakeably that of a scientist with a special interest in botany, soils and geology, but he also had a keen eye for topographical detail. During the north Devon part of his tour he noted that:

> Before we proceeded to Linton, we resolved to visit Combe-Martin, a village surrounded by lodes of iron and lead. Our road conducted us through a bold mountainous country abounding with spots most highly picturesque . . . We found Combe-Martin placed in a dale, along which it extends at least a mile from the sea-shore. The scenery of the latter is really magnificent: its more prominent parts are singularly striking, and have the happiest accompaniments imaginable. The sea enters a little cove at Combe-Martin, commodious for the mooring of small vessels; and here the produce of the mines is shipped for Wales and Bristol.[78]

But Maton is not all rosy-eyed delight. In describing Powderham, he does not hide his dissatisfaction:

> We were led to expect a noble situation for the castle, but how great was our disappointment to find it almost in a flat, very much exposed on the side towards the Channel, and with a broad marsh in front of it. It faces the river, but little pains have been taken to open the view to it with advantage, or to heighten the effect of those magnificent materials which nature has furnished.[79]

Clearly the surviving accounts of observant travellers represent an important, though frequently overlooked, contribution to the corpus of studies on the topography of the South West. Their interpretation certainly poses problems, however, and their usefulness is highly variable. But the best among them rank not only as works of literary merit, but also of genuine scholarship.

The Agricultural Topographers

Alongside the traditional topographical and historical accounts and the observations of travellers who visited the region, such as those described above, the late eighteenth and nineteenth centuries saw the emergence of 'agricultural topography' as a scholarly pursuit with a well-defined practical purpose. This was, of course, the era when the idea of 'improvement' caught the national imagination and scarcely any activity escaped the eye of expert analysts who sought to recommend the means of its advance. Although naturally focused on the farming landscape, as Sarah Wilmot's chapter clearly demonstrates, the works of agricultural writers contain much valuable eye-witness description of contemporary rural landscapes and society.

A direct link may be traced between the county surveys of agriculture promoted by the Board of Agriculture after its foundation in 1793, and taken up by the Royal Agricultural Society of England, which was established in 1838, and the deliberations and recommendations of the Georgical Committee of the Royal Society in later Stuart times. Indeed, by presenting his observations in the form of an itinerary, studies like that of William Marshall, who published the *Rural Economy of the West of England* in two volumes in 1796, may even be seen as the direct heir of Leland, Camden, Norden and Carew.[80] While Marshall's work was on a regional scale, augmented by detailed investigations in particular areas, the agriculture of each of the four South-Western counties was minutely and separately described in a series of 'general views' including that of Robert Fraser whose *General view of the agriculture of the county of Devon* was published in 1794, while John Billingsley's *General view of the agriculture of the county of Somerset* came out a year later.[81] These and a number of other important surveys of a similar character are meticulously reviewed by Sarah Wilmot who notes that the central objectives of the writers were shaped and informed by Enlightenment philosophy with its tendency to emphasise reason and 'universal' principles. They were written in an age which strongly valued scientific analysis, experiment and observation. But judgements on the aesthetics of landscape also feature prominently in these studies. In that sense, the agricultural topographers are as much a reflection of the general fashions and preoccupations of their time as the Tudor chorographers who were writing two centuries earlier. Sometimes neglected as a source of topographical information, the works of the agricultural writers are in fact a rich and rewarding seam and their studies are rightly accorded status in this volume.

Using the Works of Past Topographical Writers

In reviewing the work of a range of past topographers who have focused on the counties of South-West England, contributors to this book implicitly identify a range of questions that might usefully be posed by today's historians. Foremost amongst such questions are those which touch upon the authenticity and credibility of past studies. It is pertinent to ask how the information upon which they are based was compiled and from what sources it was drawn. Questions regarding the intellectual fashions and trends which prevailed at the time when the work was carried out are equally relevant and may shed light on its character and the motivations of the author. Thus, in Tudor times, when the county gentry sought to celebrate the uniqueness of their 'country', or home area, and the place of their family within it, antiquarians were preoccupied with the ancient origins of the community, manorial descents and the pedigrees of the important local families. By the later seventeenth century, emphasis shifted towards natural history and the description of field archaeology. And towards the end of Stuart times, travel writers such as Fiennes and Defoe sought to capture the diversity of England and, with the eye of an outsider, they made comparisons between its constituent parts.

Superimposed on the wider academic concerns of the time within which they worked, the judgements of past writers will also reflect their educational experience, the intellectual circles in which they moved, and their personal background. An example will show how important this can be. Richard Carew served as one of Cornwall's deputy lieutenants and a justice of the peace; the latter role meant he participated at the quarter sessions when matters connected with the Elizabethan poor laws were adjudicated. There can be little doubt that this experience provoked from an otherwise tolerant and sanguine man an uncharacteristically vitriolic outburst against Irish paupers in his *Survey of Cornwall*:

> We must also spare a roome in this Survey, to the poore . . . Ireland prescribeth to be the nurserie which sendeth over yeerely, yea and dayly, whole ship-loades of these crooked slips, and the dishabited townes afford them rooting: so upon the matter, the whole County maketh a contribution, to pay those Lords their rent. Manie good Statutes have beene enacted for redresse of these abuses, and upon the first publishing, heedfully and diligently put in practise: but after the nine dayes wonder expired, the law forgotten, the care abandoned, and those vermine swarme again in everie corner.[82]

Carew further betrays his elevated status as a member of the county gentry in comments on the common folk. His *Survey* contains a detailed description of Cornish mining technology where he explains the construction of addits to drain water from deep workings. Clearly impressed by their ingenuity, Carew was nevertheless surprised that such devices could be so well understood by uncouth workers. His comment would hardly pass today's exacting tests for 'political correctness':

> If you did see how aptly they cast the ground, for conveying the water, by compassings and turnings, to shunne such hils & vallies as let them, by their two much height or lownesse, you would wonder how so great skill could couch in so base a Cabbin, as their (otherwise) thicke clouded braines.[83]

In Carew's day, topographers wrote for other members of their own class and there is plentiful evidence that their manuscript as well as published works were circulated among them. As already noted, Carew of Antony was acquainted with John Hooker of Exeter and with William Camden of London. The work of Norden was widely known and used, and Camden drew heavily on the work of Leland, as Harrison had done before him.[84] Thomas Westcote and Tristram Risdon appear to have plagiarised the manuscripts of John Hooker.[85] And Risdon admired Westcote as 'a man, to give him his due praise, endowed with many good parts and lover of antiquity'.[86] Together with the fact that sources are rarely cited directly, this widespread 'sharing' of information amongst early writers raises questions about the truth of their claims that they based their work on their own eye-witness observations.

Many topographical studies are in fact highly derivative. Indeed, information gathered about the South West by John Leland in the 1530s and 1540s, and by Carew in the 1590s, was still being reproduced by writers in the eighteenth and nineteenth centuries. Modern local historians clearly need to check carefully the provenance of the testimony provided by early writers. But notwithstanding all the pitfalls and difficulties which undoubtedly face the user of past topographical studies, there can be little doubt of their value as sources from which a picture of past landscapes can be reconstructed.

Topographical Information

While space does not permit a lengthy consideration of those aspects of landscape and society that attracted the topographical writers who described the South West of their own time, a selection will serve to indicate some of the riches they contain. Descriptions of the natural environment and the physical landscape are a common ingredient and allow a glimpse not only of what the countryside was like centuries ago, but also how it was perceived by contemporaries, and the kinds of constraints which it imposed on the way of life of the population. Writers thus attribute vernacular building styles and materials to the high rainfall and strong winds which characterise the South-Western climate. They note that travel and communications were directly affected by the hilly country of the far west, and agricultural practices were directly influenced by the character of the local soils and vegetation. Early writers remark on the inexorable depletion of stocks of timber. Thus, for example, Carew's Survey of Cornwall blames the demands of ship-building and house-building which, he said, 'have bred this consumption'.[87] Risdon reports similar depletion in Devon:

> For trees, it hath the like variety as are found in other places of this Kingdom; with which in former times it hath more abounded. For now, what with good husbandry, and cleansing of the ground, and what by ill husbandry in felling and selling, the trees and timber are well shortened; which we in part feel, and future times will find wanting.[88]

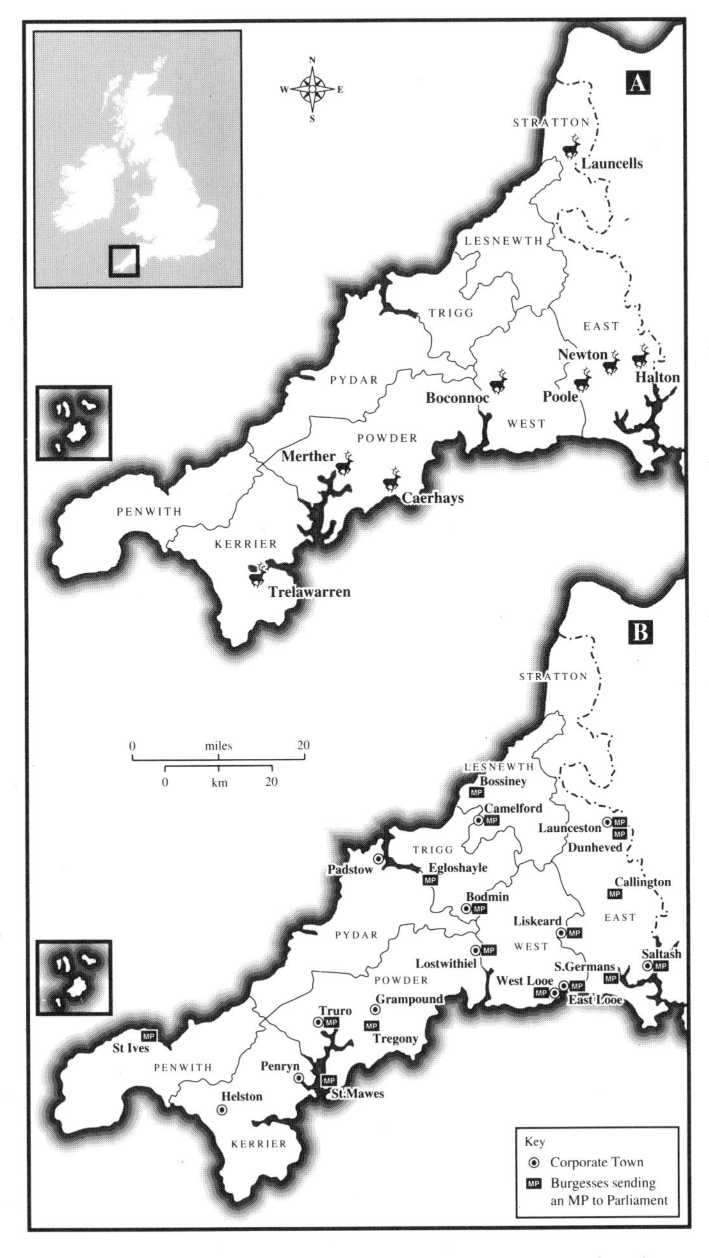

Figure 1.5 Cornwall's deer parks (map A), corporate towns and parliamentary seats (map B), recorded in Richard Carew's *Survey of Cornwall*, 1602.

Disparking was also well underway in both counties. Figure 1.5 shows how few deer parks remained in Cornwall according to Carew. Of Devon, Risdon remarks, 'howbeit those of pleasure are much diminished: for many parks are disparked, and converted from pleasure to profit; and from pasturing wild beasts, to breeding and feeding of cattle, sheep and tillage'.[89] Topographical descriptions indeed shed light on a host of themes including local agriculture, fishing, mining and quarrying. They often discuss the quality of their county's resources and the manner in which they are exploited. Descriptions of the techniques and practices customarily employed can be extremely detailed. It is, for example, Richard Carew who explains that Cornish slates were exported to Brittany and the Netherlands. He also discusses the methods by which underground tunnels were propped in the tin mines, though the methods were not entirely reliable: 'the loose earth is propped by frames of timber-worke, as they go, and yet now and then, falling downe, either presseth the poore workmen to death, or stoppeth them from returning'.[90] Carew describes how Cornish farmers converted old pasture to arable by paring, burning and dressing it with sea-sand at the rate of sixty sacks to the acre; how Devon and Somerset drovers brought their cattle to graze the summer pastures of north Cornwall; how fish was smoked and salted for export to Spain and Italy; and how the common folk amused themselves with part-singing, wrestling, hurling, or attending performances of miracle plays, or 'Guary miracles' as they were known. He reports that rats were a problem in Cornish homes, eating food, clothes and writings by day, and dancing 'their gallop gallyards in the roofe at night'.[91] He reports changing building styles where larger windows, thinner walls, glass and plaster were innovations of 'late yeere's introduction'.

The topographers describe their county's incorporated boroughs, market towns and ports; they note where the annual fairs were held; they report the locations of the quarter sessions and assize courts; they list the local militia, forts, and beacons; and they often report on the taxable wealth of the population (Figures 1.5 and 1.6). The pedigree of important families and the descent of their properties is presented, often in great detail. The locations of ancient colleges, schools, hospitals and former monasteries are almost invariably identified, as are the important landscape antiquities. Lists of parliamentary seats and ecclesiastical divisions are also usually included.

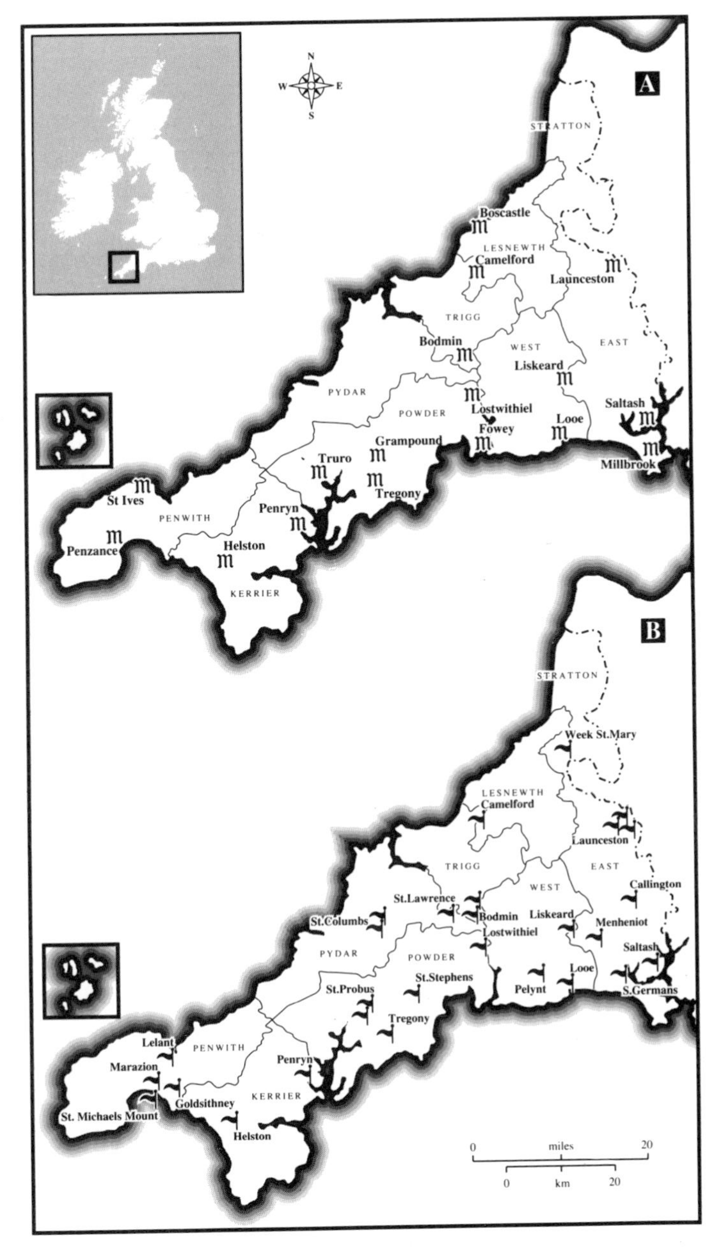

Figure 1.6 Cornwall's weekly markets (map A), and annual fairs (map B), recorded in Richard Carew's *Survey of Cornwall*, 1602.

Those authors imbued with a desire to follow an itinerary and encounter their 'country' parish-by-parish, as eye-witnesses, inevitably discuss the problems encountered by travellers. Roads, bridges, fords and ferries are thereby encompassed in their descriptions. Moreover, the advantage of choosing a locally-born horse is often spelled out. With regard to Devon, Tristram Risdon pointed out that:

> For breeding of young beasts, this county is well stored; and the like for geldings and horses, whose goodness and endurance overmatcheth many places; as appeareth by those that come from other parts, who become lame and tire in their travel; whereas ours can endure any hardships in other shires . . . This county, as it is spacious, so it is populous, and very laborious, rough, and unpleasant to strangers travelling those ways, which are cumbersome and uneven, amongst rocks and stones, painful for man and horse; as they can best witness who have made trial thereof. For be they never so well mounted upon horses out of other countries, when they have travelled one journey in these parts, they can, in respect of ease of travel, forbear a second. And therefore so much less passable for the enemy, with his troops of war, for it doth swell up with many hills, which cause as low valleys, much resembling Cornwall in form.[92]

Cornwall's horses, 'commonly bred, coursely fed, low of stature, quick in travel', were likewise reported by Carew to be 'most serviceable for a rough and hilly countrie'.[93]

Some writers also find ample space for anecdotes and asides. In this respect, Richard Carew's *Cornwall* is unsurpassed. For example, we hear of Martin Trewynard who owned a poisonous snake but had removed its fangs. Trewynard liked to frighten young women and hear them scream when he produced the snake from his coat pocket and kissed its mouth. But the snake grew a new fang and bit Trewynard on the lip, thus ending his waggish ways. Carew's sentimental tendency is revealed in the tale of the charitable dog of Saltash:

> And if *Plyny* now lived, I suppose he would affoord a roome, in his natural History, to a dogge of this town, who (as I have learned by the faithfull report of master *Thomas Parkins*) used daily to fetch meate at

> his house there, and to carry the same unto a blinde mastiffe, that lay in a brake without the towne: yea, (that more is) hee would upon Sundayes conduct him thither to dynner, and, the meale ended, guide him back to his couch and covert againe.[94]

This bizarre story clearly struck a cord: it was reproduced by several later writers including Daniel Defoe. Carew's splendid prose thus immortalised both the commonplace and the curious. His work sparkles across the centuries.

Although other South-West topographers do not match the peerless Carew, taken together, their eclectic, whimsical, sometimes cryptic, always devoted studies, which reveal so much about the topography of Cornwall, Devon, Somerset and Dorset, are rarely dull and often offer unalloyed delight to the modern researcher. This modest book is offered in celebration of their work.

2

JOHN LELAND IN THE WEST COUNTRY

JOHN CHANDLER

Four centuries of topographical writing have mined a rich seam in the works of the apparently ubiquitous Leland. It is indeed rare, whenever the Tudor history of a town is under scrutiny, not to find a guest appearance from the man scholars used to refer to in reverential tones as 'the king's antiquary'. But despite his enduring popularity as a source, Leland has not received from historians the critical attention that his importance warrants. In fact, rather more consideration has been paid in recent decades to his achievements in other fields, as biographer, bibliographer, and poet. In this chapter I should like to examine how Leland's knowledge of the West Country was acquired, and thereby to make an assessment of his value to students of West-Country history.

The salient features of John Leland's life, so far as they are known, may be briefly stated.[1] He was born in London in about 1503, and was enabled through patronage (despite being orphaned) to attend St Paul's School, before university study at Cambridge, Oxford and Paris. In the course of this education he made influential friends, acquired a wide range of humanist interests, including poetry and historiography, and developed a fascination for medieval culture and history. He taught, took holy orders so as to gain sinecure preferments, and in 1530 was appointed to a post in the royal libraries. Three years later (and perhaps at his own suggestion) Henry VIII commissioned him to inspect monastic libraries and make lists of important contents, a task which appears to have occupied him from 1533 until the dissolution was complete, in 1539/40 (Figure 2.1).

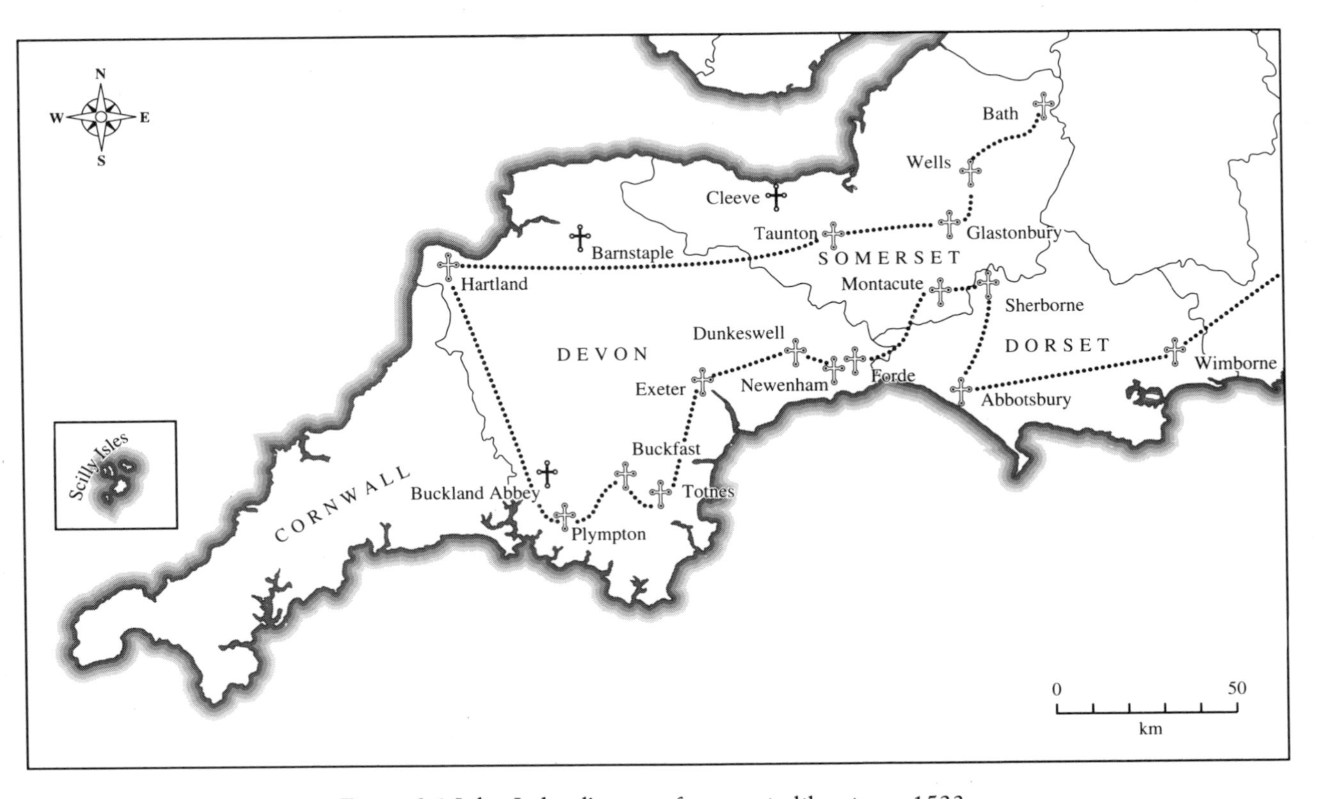

Figure 2.1 John Leland's tour of monastic libraries, *c.*1533.

During these years he continued to write poetry, and developed a taste for what may best be described as sight-seeing; with the result that from c.1539–45 he supported himself on a series of journeys in England and Wales which he called (and we call) his itineraries. At the end of 1545 he reported to Henry that he had collected sufficient material to write an ambitious series of books, on national biography, literature and topography, and to furnish information for a map of the kingdom. But in 1547 (the year of Henry's death) he became incurably insane, and he died in 1552. In his lifetime a few poems and pamphlets appeared in print, but none of the major projects was accomplished.

After his death the mass of notes remained in manuscript, and was passed between antiquaries, some of whom copied and/or lost portions. In 1632 most of what survived was presented to the Bodleian Library, Oxford, where it remains. Much of this material was published during the eighteenth century,[2] and the 'Itinerary', with which local historians are principally concerned, was scrupulously edited anew by Lucy Toulmin Smith and published between 1906 and 1910 (reprinted in 1964).[3] This remains the standard work, although many West-Country historians may be more familiar with an inferior edition by Pearse Chope, published in his *Early Tours in Devon and Cornwall* in 1918 (reprinted in 1967).[4] In recent years considerable attention has been paid (by James Carley, Caroline Brett and others[5]) to Leland's poetry and library lists, as well as to his work on, and his own place in, English historiography. The present author has published a popular version, in modern English, of much of the 'Itinerary'.[6]

Leland explored the West Country twice on his itineraries, in 1542 and (probably) 1544 or 1545 (Figure 2.2). One or more earlier tours had taken in West-Country monasteries, and it may have been during such a tour that he wrote a description of Cornwall. He also wrote poems which include topographical details about West-Country places, and there are stray references elsewhere in his works. We shall consider this material, so far as possible, in chronological order.

The library lists, preserved in the Bodleian Library and published in the eighteenth century as part of Leland's *Collectanea*, display geographical sequences which suggest the journeys he took in order to compile them. One sequence implies a route beginning and ending in London which took in the West Country, the South Midlands and East

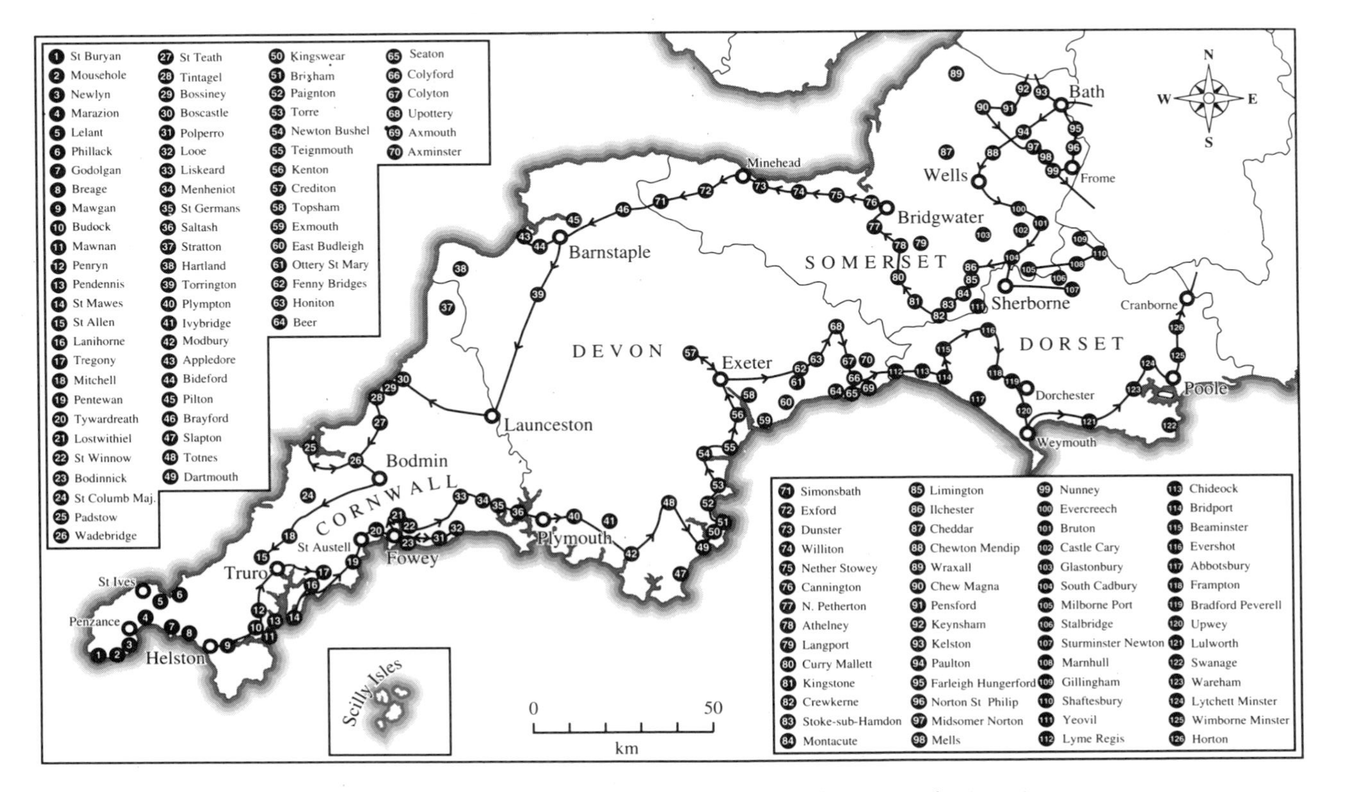

Figure 2.2 John Leland in South-West England, 1542 and 1544–5.

Anglia.[7] Leland's manuscript is neatly written, as if a fair copy, with gaps left against certain monasteries for later revision, and it leads us from Guildford and Waverley (near Farnham) to the south coast, to Wimborne, Abbotsbury, Sherborne, Montacute, Forde, Newenham, Dunkeswell, Exeter, Totnes, Buckfast, Plympton, Tavistock, Hartland, Taunton, Glastonbury, Wells and Bath, before moving on to the Cotswolds and beyond. Gaps are left against Buckland, Barnstaple and Cleeve, and brief historical or topographical notes, rather than library books, are all that are entered against some other houses.

If this sequence does indeed represent a single journey, it is possible that the West-Country portion took place in 1533, the first year of Leland's commission. This is because during his 1542 itinerary he recalled a previous visit to Bath, made nine years earlier.[8] His interest in West-Country literature, however, can be traced back even further, to his student days in Paris, when he befriended a professor of rhetoric in order to search his library for the works of Joseph of Exeter.[9]

Apart from his bibliographical quest, two other of Leland's later concerns are already apparent during this hypothetical 1533 journey. One was note-taking about the foundation, situation and history of the monastic houses he visited. At Montacute, for example, he catalogued one book, Paschasius' *De septem sacramentis*, but also offered (in Latin) this confused explanation of the place: 'Mons Acutus, in the British tongue Brent Cnolle, which means the mountain or hill of frogs. Mons Acutus takes its name from the natural feature [the pointed hill]. The monastery's founder was the Count of Mortain, who is buried at Bermondsey.'[10] His other concern was with the authenticity of Geoffrey of Monmouth's account of Arthur. At Glastonbury Leland recorded the date of the discovery of Arthur's remains, and an epitaph to him. Three years later, in 1536, he pulled together into a pamphlet, *Codrus*, all the evidence, from Glastonbury and elsewhere, which he had so far discovered in support of the historical Arthur, in order to attack the sceptical views of Polydore Vergil.[11]

If we may trust the library lists, this early journey did not extend into Cornwall. But it is clear that by 1536 he had explored the Duchy, because there exists among the notes in the *Collectanea* a systematic description of Cornwall, similar to those for some other counties (Cheshire, Herefordshire, Lincolnshire, etc.), which cannot be regarded as part of the later itineraries. This account of Cornwall refers in the present tense to two monasteries, Tywardreath and St Carrok, which were both dissolved in 1536.[12] Unlike the 1542 itinerary, in the course of which Leland appears to have travelled no further west than St Ives and Godolphin (near Helston), this earlier description includes eye-witness accounts of Newlyn, St Buryan, Mousehole and other places in West Penwith.

It is not as yet possible to date most of Leland's movements during the 1530s,[13] nor is the sequence of the later itineraries certain. The autograph, so far as I am aware, gives only one indication of date, 5 May 1542; but that is important, because it heads the principal West-Country tour, and appears to be the date of his setting off from London. His journey ended at Winchester, presumably later in the same year, because by 24 March 1543 (the last day of 1542 old style) he had composed and seen published in London an elegy of some 180 lines in honour of his friend, the poet Sir Thomas Wyatt.[14] Wyatt had died in October 1542, paradoxically while he himself was on a journey to the West Country. He had been dispatched from London in haste to meet a Spanish emissary arriving at Falmouth, but over-exerted himself on the journey and died at Sir John Horsey's house, Clifton Maybank near Sherborne, which Leland had visited a few months earlier.

Leland's 1542 itinerary took him from London to Great Haseley near Oxford (to the living of which he had recently been presented), then via the Cotswolds to Wiltshire and Bath, down through Somerset to Sherborne, then skirting the Somerset Levels to Bridgwater and the coast, across Exmoor to north Devon and into Cornwall at Launceston. After touching the north coast around Tintagel he set off through Cornwall to the Godolphin seat near Helston, returning along the south coast via Fowey to Plymouth. Thereafter the journey took him to Totnes and Exeter, Mohun's Ottery (a seat of the Carews), and through Dorset to Salisbury and Winchester. Here, after an excursion to Portsmouth and Titchfield, the itinerary concludes.[15]

The account of this journey is the most valuable by far of Leland's legacies to the West-Country historian, and we shall examine aspects of it in greater detail shortly. But in addition to its place within his long-term scheme of nationwide note-taking for the ambitious undertakings which he was never to fulfil, the journey also furnished him with important evidence for projects current in the period 1542–4. One of these was a reworking of his defence of the historical Arthur. The 1542 journey took in not only Glastonbury, but also South Cadbury, where he saw for himself, 'Camallate, sumtyme a famose toun or castelle, apon a very torre or hille, wunderfully enstrengtheid of nature . . .'. Roman coins and other antiquities had been found there, including a silver horseshoe. 'The people', he noted, 'can telle nothing ther but that they have hard say that Arture much resortid to Camalat.'[16] In 1544 Leland published (in Latin) a much augmented version of *Codrus*, which he called *Assertio incomparabilis Arturii*; it includes eye-witness descriptions of Arthurian sites, including South Cadbury. 'Good gods, how very deep are the ditches here! How many the earthen ramparts piled up! And what sheer drops there are! In a word, it seems to me an absolute miracle, both natural and man-made.'[17]

The *Assertio* is a polemical work in Latin prose, designed to defend and strengthen the Tudor claim to the throne. Leland the loyal courtier is seen also in another work in progress at the time of the 1542 journey. This was a poem of nearly 800 lines celebrating the birth of Henry's son, the later Edward VI, and was perhaps begun amid the euphoria surrounding the birth in 1537. Leland's *Genethliàcon illustrissimi Eaduerdi* . . . was not completed and published, however, until 1543, presumably in time for the child's sixth birthday on 12 October.[18] Three titles (Prince of Wales, Earl of Chester, and Duke of Cornwall) had been conferred on Edward at birth, and much of the poem is devoted to describing the supposed birthday celebrations in these three regions. Leland had visited Wales and Cheshire, probably during the period 1536–9, and his return to Cornwall in 1542 perhaps prompted him to complete the work.

It is a curious fact about the 'Itinerary' that in its pages Leland the English topographer almost never becomes Leland the Latin poet.[19] The reverse, however, is not true. In the *Genethliacon*, as in a later poem, *Cygnea Cantio* (1545), which describes a swan's journey down the Thames, the poet delights in displaying his topographical erudition. Time and again miniature geography lessons punctuate the flow, and to dispel any uncertainties both poems are followed by prose commentaries on all the people and places mentioned. The commentary to *Cygnea Cantio* occupies seventy-four pages, and is a substantial corpus of topographical information in its own right (which has never, incidentally, been made available in English translation). It is out of all proportion to the poem itself, which takes up fewer than sixteen pages.[20] The commentary, or Syllabus as it is entitled, to *Genethliacon* is more modest. An English rendering of both the section of the poem relating to Cornwall, and the Cornish entries in the Syllabus, is appended to this paper.

One further foray into the West Country took place after the 1542 itinerary. Beginning from Oxford, Leland travelled through Wiltshire to the Bath and Bristol region, exploring in some detail places in south Gloucestershire and north Somerset, and finishing around Blackmore Vale, on the Somerset–Dorset border.[21] His destination was probably West Knoyle, near Mere (Wiltshire), where he held a prebend. As one might expect, the journey seems to have been planned so as not to duplicate the earlier route, but to explore towns (most notably Bristol) which Leland had missed in 1542. In common with all except the 1542 West-Country itinerary, this journey is undated. Leland's autograph, in fact, is lost, and our knowledge of it is derived from a sixteenth-century copy, probably after the original had become defective.[22] It post-dates the 1542 journey, and internal evidence suggests 1544 or 1545.[23] It comes, at any rate, towards the end of Leland's working life, and almost certainly describes the last occasion on which he visited the West Country.

Enough has probably been written to demonstrate that there is more for the West-Country historian in Leland's output than will be discovered simply by looking in Chope's or Toulmin Smith's editions of the 'Itinerary'. The library tours, the set-piece poems and the commentaries appended to them, are all of topographical interest, to say nothing of Leland's transcripts of extracts from medieval chronicles and histories, some otherwise lost. Then there are his biographical collections on British authors, published in 1709 as the *Commentarii . . .*, which include many West-Country figures. His biography of Boniface, for example, includes a discussion of the etymology of the name Exeter.[24] His shorter poems, too, unpublished during his lifetime, include encomia to West-Country contemporaries, such as John Clerk, Bishop of Bath and Wells, 1523–41 (written, presumably, with an eye on clerical preferment). There is also a poem of forty lines in praise of Bath.[25]

Nevertheless, it is the 'Itinerary', and in particular Leland's account of his 1542 journey, which is most often quoted or cited in histories of West-Country places. Many authors find the allure of Leland's Tudor English irresistible, and include long verbatim quotations—harmless enough, perhaps, although such a practice carries certain risks. First, in fairness to Leland, it should be pointed out that, unlike Camden, Defoe and other topographers, only his notebooks exist, and quotations from the 'Itinerary' do not reflect the polished style, in Latin or English, of which he was quite capable. Second, in fairness to the reader, Leland's sixteenth-century usage of certain key words, such as town or mile, should be clarified. Third, users of Leland should try, wherever possible, to identify his sources, and therefore his credibility.

Leland was a careful and critical researcher, questioning the accuracy of his sources, seeking corroboration wherever possible, citing authorities and conducting his own fieldwork. Anyone who wishes to see Leland at work could do no better than browse among the notes concerning various places printed in part 11 of Toulmin Smith's edition of the 'Itinerary'.[26] Here in close proximity will be found phrases such as: 'Averey parson of Dene tolde me that he had redd that . . .; But loke bettar for this mattar. Sens he tolde me that it is in S. Kenelme's lyfe that . . .'. Or again: 'As far as I could perceyve by questioninge with the auncient servaunts and officers of the Marquese of Dorsete . . .'. If he discovered that he had been misinformed, he wrote himself a correction: 'Where as I have wrytten that the castell and vyllage of Stoke Fleminge stode at Dertmowthe, I made ii errors . . .'. Some mistakes, of course, went undetected, such as his note, 'At Bridporth be made good daggers'. Leland was perhaps unaware of the town's rope-making industry, and it was left to Fuller a century later to explain that 'a Bridport dagger' was a hangman's noose.[27]

Careful reading of the 1542 itinerary will show that it is a blend of personal observation (including scrutiny of inscriptions), answers to questions which Leland asked his informants, and information from various written sources, including books and documents which Leland's hosts along the way had in their possession. While engaged as the king's official on his library tours during the 1530s he was, of course, able to enjoy monastic hospitality from the heads of the religious houses which he visited; and he gathered information from conversations with them, as well as from the books in their libraries. His recollection in 1542, for example, of the previous visit to Bath, seems to imply that he was shown around the priory buildings by the prior himself, whom he questioned about their history.[28] But in 1542 the monks had gone, and Leland had to rely on the web of acquaintances which he had cultivated during his academic career, through his sycophantic poetry, and at court, in order to provide him with hospitality along the way, and to open the doors and the muniments of the great gentry houses. In Cornwall the Godolphin, Arundell and Treffry families were among his hosts, and in Devon Sir Philip Champernowne of Modbury and Sir George Carew of Mohun's Ottery near Honiton.

Thomas Treffry, Leland's host and informant at Fowey, was engaged by the king in overseeing the construction of coastal defences, including the castle at St Mawes. His encounter with Leland resulted in a commission to the latter to compose Latin verses in praise of Henry as an inscription on St Mawes Castle.[29] As part of the defensive scheme, the south coast of Devon and Cornwall was surveyed, and a chart in four sheets was produced. It seems possible that Leland was shown the sheet relating to Falmouth harbour by Treffry during his visit. This would explain an otherwise puzzling reference to St Mawes: 'Half a mile from the hedde of this downward to the haven is a creke in manner of a poole with a round marke made in the charte on the which is a mille grinding with the tyde'.[30]

Leland's visit in 1542 to Sir George Carew (who was destined, as commander of the ill-fated *Mary Rose*, to perish when it sank in Portsmouth Harbour three years later[31]) provides a good illustration of his methods. He is quite explicit about his route on leaving Exeter (Clyst Bridge, i.e. Clyst Honiton; Taleford, i.e. Fairmile; Fenny Bridges; Honiton; Mohun's Ottery, between Luppitt and Monkton). Mohun's Ottery was Carew's home, and after Leland had visited him the itinerary resumes, to Colyton, Colyford, Seaton and Axmouth. While a guest at Mohun's Ottery, Leland collected information which included the following: details of the course of the River Otter above the house; an explanation of the link between the Mohun family, the Carew family, and the house; and the course of the Otter from the house down to the sea, including descriptions of Otterton, East

Budleigh, and what is now Budleigh Salterton.[32] He is explicit that the genealogical information came from Sir George himself, and we may assume that the description of Otterton, as 'a praty fischar toun', also derived from Carew, not from Leland's own observation. What is quite clear is that Leland himself did not (at least not on this occasion) visit the settlements at the mouth of the Otter, and whatever he wrote about them must be presumed to be hearsay.[33] It therefore behoves anyone citing Leland to try to discover from the context whether the passage is an eye-witness account or a second-hand report.

Leland's visit to Mohun's Ottery illustrates another facet of his method. In the autograph the last three lines of one folio are repeated in error on the next, and then crossed through.[34] The mistake suggests that in this instance the autograph is Leland's fair copy of an earlier draft, and such an assumption can be strengthened by examining other portions of the 1542 itinerary. The autograph of the journey is contained in two notebooks, and at the end of the first are two folios of notes about the Bodmin area. A fuller account of the same leg of the journey opens the second notebook. At the end of this second book are no fewer than fourteen folios of notes about West-Country places, including transcripts of passages from books and inscriptions, which correspond to material in the more detailed narrative of the journey which precedes them.[35]

If, as these duplications suggest, Leland was in the habit of copying and amplifying rough notes, which he did not then discard, this may help to explain why some of his itinerary accounts are more detailed than others, and why some (relating to Kent for example, and Wales) appear not to follow a logical geographical progression. The sad tale of neglect and damage which his manuscripts suffered after his insanity and death is best summed up by a comment of William Harrison, who saw them some twenty-five years later, in *c.*1576: 'books utterly mangled, defaced with wet and weather, and finally, unperfect through want of sundry volumes'.[36] A likely consequence of such treatment would be that some of the survivors are the rough notes, while others are the fair copies. It is thus a bonus for West-Country historians that, for parts of the 1542 tour, both survive.

How much information did Leland record about the West Country which has not survived? Although not the first topographer to write itineraries (that honour goes to William Worcestre[37]), Leland's work has been regarded for centuries as seminal to the progress of English topographical and local history studies. His reputation in this regard shows no sign of diminishing.[38] The principal reason for this status is that everyone who came after him knew, used, and quoted from his work, frequently without due acknowledgement. How much 'lost' Leland is embedded anonymously in the works of, for example, Camden and Stow is an unanswerable question.

But it is important that users of Leland should recognise that the surviving portion of the 'Itinerary' is not the limit of the information which he has to impart. I have tried in this chapter to show that prospectors will find it worthwhile to look beyond the 'Itinerary', and quarry elsewhere in the surviving corpus of Leland's works. A second approach, which must await another occasion, would be to try to discover who, among the historians and topographers of the generation after Leland, had access to now lost manuscripts.

William Harrison, who contributed descriptions of England and Britain to Holinshed's *Chronicles*, first published in 1577, borrowed some of Leland's manuscripts from John Stow, but as Stow's own transcripts of those portions of the 'Itinerary' which he posessed have survived, this does not help us. However, by the time that Harrison came to prepare another edition of his work, published in 1587, he had seen more. In his new edition he referred to a work which he called Leland's *Commentaries*, book 13, and he appears to have cited passages about East Anglia which are not among Leland's surviving manuscripts.[39] Stow referred to the 'Itinerary' as *Commentaries*, but he only transcribed ten volumes.

If, by close comparison of Harrison's first and second editions, hitherto unrecognised passages of Leland might be recovered, it is most likely that they would describe places in East Anglia and the South-East, for which itineraries do not survive. The local historian must ever be grateful to Leland that he wrote so much about the West Country, and to his successors that so much of it has been preserved. If yet more remains to be discovered or identified, then that would be a bonus indeed.[40]

Appendix

In this appendix, I offer an English translation of part of Leland's *Genethliacon illustrissimi Eaduerdi* (1543), or birthday ode for Prince Edward, which he probably completed after his 1542 West-Country journey. The Cornish section, here translated, occupies lines 500–605 (pages XIX–XXII) in Hearne's edition (see note 18 below). For completeness I append a translation of the Cornish entries from the Syllabus which follows the poem (this, inexplicably, includes no Cornish material after the letter F). I am not aware that any of this material has been presented in English before.

The Cornish uplands face the setting sun. It is a wide country indeed, past which the waters of the River Tamar flow down from the north, and set their fleeting course southerly. But further still is this fortunate region's uninterrupted length, extending like a narrow cone to its point. Here on one side it touches Marazion (*Forum Jovis*), and on the other tastes the ebbing tides of the horned River Hayle (*Alaunus*). But then it grows, spreading broadly about a glittering ring, and suddenly becomes a lofty peninsula of new land, whose two headlands project into the sea. Here the vigorous news has come on agile wings, and with overwhelming joy has given loud and clear the signal to proclaim with grateful voice the birth of a prince. Hearing the news the sturdiest crowd of Cornishmen swarm up from the mines, and from the fire-spewing furnaces too. The sailor returns cheerfully from the sea, the merchant seeks out the most prominent places in the busy market-place; and each in their turn make carnival.

A fair spot clothed in verdant grass is surrounded, crown-like, by a vast throng of people. In their midst the naked bodies of two wrestlers are locked in combat, as sweat-covered they try their strength. The champion stands firm, solid as oak. His opponent falls to the ground on his back. The shouting reaches the vault of heaven as the vanquished man is removed. The victor celebrates, and his victim nurses injured pride. Each takes due reward for the splendid victory. Those defeated at wrestling demand to compete in the racing. An ashen post is set up at the furthest point of the course, and both runners strive to touch it first with their hand. You would imagine that the speed of their flight might outstrip deer. This time fortune smiles on the vanquished, and victory adjudges the result to be equally balanced. Eventually the inhabitants make vows to their own beloved duke, after this fashion: 'Illustrious Edward, we pray that you may live a long life on earth, and that you may have the fortune to exceed in whatever manner you wish even the famous deeds of your ancestors'.

In Falmouth harbour (*portus Falensis*) nature has amused herself with a manifold refuge, bestowing on fleets unknown a generous lodging whenever Aeolus rages with his devouring southerly wind. Here the sailors have laid aside their boats, and have organized their own contests. Their captain has set the time-honoured rules, and poles have been fixed as markers on the sodden sand. Next the two contestants prepare themselves for the noble task. Then mighty boatswains sounded the signal on the golden pipe, and the whole bay of the delightful harbour echoes to its sound. At this sign they strain powerfully on the oars, and by turns one boat then the other is in front. When one is seen to be far ahead of the rest, so that it might be awarded the victorious accolade of Duke Edward, then the boatswains make another blast on the golden pipe. The wave, churned up by the fir-wood oars, crashes with a deeper note, and the sailors reply, chanting these words: 'May our Duke Edward flourish, foremost glory of the Cornish race, our light, our love, and our honour'.

Then throughout the six towns which Cornwall nurtures with her plentiful bosom, a vast throng of merchants affirmed the special honour with glittering processions. Launceston (*Dunevetum*), more eminent than Stephen's sanctuary, was one, looking down on the others from the rocky summit of its cliff, and taking pride of place with deserved glory. Bodmin (*Bosuenna*), that busy place, came next. Next followed Liskeard (*Lesceretum*), and Lostwithiel (*Losuitiellum*), the famous house where tin is stamped. Truro proved itself an equal companion to the others. Lastly Helston (*Aluania*) echoed with applause. With one expression and with one voice the thankful mouths of all the citizens uttered words like these:

'Salvation and greatest hope of our country, Duke Edward, may you live long, and may the best of fortune favour your life. As maturer years bring you to strong manhood may you look with indulgence on our voice. Then shall we be desparately anxious to see the lord we love in person, and to hear for ourselves the words you speak. That day of days will restore our fortunes. Most illustrious lord, accept the gifts which we offer.

'Here every race of birds makes its nests on our wild cliffs. What diversity of fish, delicious to the taste, play in the waters around our coast? How great a quantity of apples hang from our sagging trees? Cattle, flocks, almost too many to count. But now we must speak of greater matters. A mighty wealth of tin flows from our furnaces, and along our cavernous coast has been discovered a seam of copper, and silver, even—if experts are to be believed—gold. Valuable pearls are our shining gifts, and the steely metal which we have found sprinkles its stars far and wide. And if you should be seized by the desire for an even greater crowning triumph, to be given you over your vanquished foe on the azure sea, then the best equipped fleet would be made up of our ships—invincible, such a fleet will secure you overwhelming victory. All these things are yours absolutely, most illustrious duke, and more besides.

'We beg of you to hold such thoughts in your memory, so that, when you have grown up to maturer years, you will honour us by visiting your servants' abodes. King Aethelstan favoured us most generously with outstanding support. His help was matched by that of the magnificent Richard, Henry's brother, who as his reward took for himself the throne of mighty Rome. It remains for us to express, with due respect, one further hope for the future, that you, Edward, may follow in your father's illustrious course of virtue. He has provided you with demonstrations of uncommon renown. Opulent Pendennis stands as one example, and most impregnable St Mawes (*Fanum Mauditi*) as another; here forts have now been built at the Fal's mouth, which with flashing thunder and menacing rock can exclude pirates, that wicked, bloodthirsty race of men'.

At these words the throng of citizens fell silent.

Alaunus: a river on the north side of the Cornish isthmus.
Alaunia: a town, in the British tongue Hellas, or Heilstoune, so called, I assume, from the river of the same name nearby. It lies next to a port commonly pronounced Heilforde.
Bosuenna: remembered as the most famous market in the whole of

Cornwall. It is now called by its more recent name, Bodmine. It is certainly a famous place, not only for its monument to Petrock of Britain, who was once esteemed as the most distinguished of holy men; but especially for the benefactions of Aethelstan king of the Anglo-Saxons, that most fortunate of conquerors, who regained control of the entire province of Cornwall. Adelstow (that is, 'the place of Adelstan'), which is a town very well known to fishermen, in common parlance Padestow, is an obvious and clear proof of his victory.

Corinia: so-called from Duke Corinus, according to the writers of the British history. There are those, such as Asser and others, who call this region Cornubia. Some call it Cornugallia, including the famous William of Malmesbury, foremost of history-writers, and others whom Polydore cites. In his history Tilbirius Anglus names the land Cornubria, as being the horn (cornu) of Britain. But I, if I might be permitted to venture my own opinion, would declare that it should be spoken as Cornewalle (that is Coriniorum Galliam, 'Gaul of the Corini', or Corinienses Wallos, 'the Welsh Corinians'). Of the Welsh we spoke a little earlier.

Dunevetum: In the British tongue Duneveth, the first city of all Cornwall, and built on a very high hill. More recently it has been called Lanstuphandune, which means 'the shrine of Stephen on the hill'. From this city originated the family of Dunevet, [i.e. Knyvett] which is very highly regarded among the people of Norfolk (apud Icenos, sive Volcas boreales). Surely Sir Thomas Knyvett was without equal, a second Ajax, upon whose death in a great sea-battle the whole of England mourned. From him sprang another most noble knight, Henry, who is also a very great patron of literature. The family name was long ago shortened in common parlance, to Nevet in place of Dunevet.

Fanum Mauditi: In the British tongue Saincte Maws.

Falensis: a port, commonly Falemuthe haven.

Forum Jovis: In the British tongue Merkiu, a coastal town, by the passage to St Michael's Mount (in insulam Michaeliniam).

3
Some Early Topographers of Devon and Cornwall

JOYCE YOUINGS

The Carew Poles

To stand in the entrance hall of Antony House, a few miles into Cornwall from the Torpoint Ferry, is to be at the epicentre, not only of the South-Western peninsula, but also of its early topographical literature. From the wall there gaze down the sharp eyes of Richard Carew (1555–1620), arguably the most distinguished member, certainly as a scholar, of that far-flung and remarkable West-Country family. His *Survey of Cornwall* must, I admit with some regret as a Devonian, take precedence of all its competitors for our attention, if only because it was the only one of the South West's early topographical works to be published in its author's own lifetime and therefore to have been published as its author left it. Antony House contains not only Carew's portrait, but also that of Sir William Pole of Shute in east Devon (1561–1635), an altogether chubbier face, not unlike that of his fellow Devonian, Sir Francis Drake. Pole's portrait came to Antony, together with quantities of his papers, following the two families becoming united by marriage. What an inheritance!

Richard Carew

But our first concern must be with the great Cornishman. Richard Carew was at Christ Church, Oxford in the late 1560s with both Sir Philip

Sidney and, perhaps more important in this context, with William Camden, and there, and subsequently at the Middle Temple, he rubbed shoulders with the cream of early Elizabethan scholars and writers, acquiring thereby both a facility for writing elegant prose and a well-tuned and inquiring mind. This made him welcome in the London scholarly circle from which emerged, in about 1586, the Society of Antiquaries. What a time to be indulging a love of antiquity! He might also have become a fashionable poet, but for the rest of his life Carew preferred to spend most of his energies living the life of a fairly well-to-do country gentleman, engaging in the multifarious administrative duties expected of him, though, as his son later recalled, when riding around the county he always carried close to him one or other of his books. From time to time he would go to London to forgather with the intellectual elite, reading the occasional scholarly paper. In this Cornwall was, and is, most fortunate, for as a result he was to become more than a mere antiquary concerned with pedigrees and manorial descents (though he was that too), but one able to survey the current economy of Elizabethan Cornwall and (although possessing no great talent as an historian) with the instinct to survey the contemporary scene against what little he could find out about the recent, and occasionally the remoter, past. And he did not hesitate to draw lessons therefrom, lessons which, though probably barely comprehensible to his neighbours and tenants, will speak volumes to modern readers with a mere smattering of economics.

One example will have to suffice. It concerned the Tudor problem of inflation. Referring to the 'dearth', i.e. high price, of corn he expressed the opinion that this was 'no way prejudicial to the good of the country', for it encouraged the farmers to grow more and hence to employ more labour. This in turn 'setteth the artificer [craftsman] on work, who were better to buy dear bread . . . which he countervaileth again by raising the price of his ware than to sit idly knocking his heels against the wall'.[1] This, of course, was the voice of the landlord, but on the whole (and he has a lot more about the Cornish economy) Carew was more thoughtful and much less complacent than his near-contemporaries in Devon. Carew, incidentally, offered as one explanation of the recent growth in population something which has only been rediscovered in our own day by the Cambridge technocrats, that the county was seeing 'younger marriages than of old'.[2]

While space does not permit a full evaluation of Carew's remarkable *Survey*, it may be noted that it reveals much about Elizabethan society that was not peculiar to Cornwall. Some two-thirds of the text consists of general description, only a minor part being devoted to the obligatory itinerary, which he dealt with hundred by hundred. Hence there is not the temptation, as with most other texts, to dip into it only for one's own parish. For those without access to the first edition of 1602, or to the microfilm copy (Figure 3.1), there is an excellent recent edition, albeit in modern spelling and with a certain amount of cutting, but otherwise a very faithful rendering of the original.[3]

John Hooker

Without admitting any prejudice as a Devonian, I must confess to an affection and indeed a great respect for John Hooker of Exeter. Like Carew he was very much a man of the world of his time, which in Hooker's case stretched well back into the first half of the sixteenth century and he was already a mature family man of nearly 35 at the beginning of the Queen's reign in 1558. Born of an established civic dynasty which, like many of Exeter's mercantile elite, claimed, in this case quite rightly, to come from gentle stock, from 1555 until his death in 1601 Hooker held the new office of chamberlain of the city of Exeter. He made it that of city manager. It was in fact a remarkable appointment by a Marian mayor, for Hooker was, in the almost literal sense of the term, one of the 'hotter sort of Protestants'. He too had been at Oxford where he read Roman law, before spending a short period on the Continent sitting at the feet of leading Protestant divines, and on his return home in 1549 he looked cut out for the heights of contemporary European scholarship. Instead he was to combine his civic duties with what today would be called lay preaching, explaining his academic backsliding to his being lured into matrimony.[4] Even so his undoubted talents as a literary man led to his writing and publishing a series of useful treatises on local government and he was also very prolific as a translator and editor. His blow by blow account of the siege of Exeter in 1549, which he himself experienced at first hand, was journalism of the best kind except for his unconcealed religious sympathies and his obsessive loathing of popular

The survey of Cornwall.

hils. The Westerne are better trauaileable, as lesse subiect to these discommodities: generally, the statute 18. *Eliz.* for their amendement, is reasonably wel executed.

Bridges. Bridges, the riuer *Tamer* hath *Polston*, *Gresham*, *Horse*, and *New Bridge*. *Lyner*, that at *Noddetor*, *Seton*, and *Loo*, two bridges of the same name. *Foy* riuer, *Reprin*, *Lostwithiel*, *S. Nighton*, or *Niot*. *Fala* riuer, *Grampord*, *Tregny*. *Loo* riuer, *Helston*. On the North coast, vpon *Camel*, *Wade*, *Dilland* & *Helland*. Vpon *Deuon*, *Trywartheuy*, *&c.* for they are worth no curious enquiry.

Traffike markets. For maintenance of traffike by buying and selling, there are weekely markets kept: In the Hundred of East, at *Saltash*, *Launceston*, and *Milbrook*. In West H. at *Loo*, and *Liskerd*. In *Stratton* H. at the Towne of the same name. In *Lesnewith* H. at *Bottreaux* Castle, and *Camelford*. In *Powder* H. at *Foy*, *Lostwithiel*, *Grampord*, *Tregny*, and *Truro*. In *Trig* H. at *Bodmin*. In *Kerier* Hun. at *Helston*, and *Perin*. And in *Penwith* Hundred, at *Pensants*, and at *S. Ies*. Of these, *Bodmyn* and *Launceston* are the greatest: this as placed in the broadest, that in the middle part of the Countie.

Faires. Fayres there are many, some which here ensue.

March 13 at *Bodmyn*, *Helston*, S. *Michaels mount*.
April 24. at *Loo*. 25. at *S. Columbs*, *S. Probus*.
May 1. at *Launceston*, *Perin*.
Iune 11. at *Mimbinet* 24. at *Lãucestõ*, *pelint*, *probus*, *Colombs*
Iuly, on *S. Margets* day, at *S. Stephens*. *S. Thomas transl.* at *Camelford*.
On S. *Iames* day, at *Golsinni*, *Saltash*.
August 1. at S *Germaines*.
On S. *Laurence* day, at S. *Laurence*.
On the Assumption of our Lady, at *Lalant*.

Sep-

The first Booke. 54

September, on S. *Mathews* day, at *Liskerd*, on S. *Bartholmews*, at *Lostwithiel*, on the Natiuitie of our Lady, at *Kellington*, S. *Marie weeke*, and *Marcasiow*.
October, on S. *Dionise* day, at *Treuenna* in *Tintagel*.
Nouember, on S *Katherins* day, at *S. Thomas*.
On S *Leonards* day, at *Launceston* and *Tregny*.
December, on *S. Nicholas* day, at *Bodmyn*.

Waights and measures. And because traffike cannot bee exercised without waights and measures, a word or two of them.

Touching wayghts, the statute 12. H. 7. which made a generall ordinance therein, did specially exempt those appertayning to the cunnage, in *Deuon* and *Cornwall*, *viz.* that they should be priuiledged to continue their former vsage.

In measures the Shire varieth, not only from others, but also in it selfe: for they haue a land measure, and a water-measure: the water-measure, of things sold at the ships side (as salt and peason) by the Inhabitants, is sixteene gallons the bushell; by strangers, betweene 18. and 24. The land-measure differeth in diuers places, from 18. to 24. gallons the bushell, being least in the East parts, and increasing to the Westwards, where they measure Oates by the hogshead.

The Iustices of peace haue oftentimes indeuoured to reduce this variance to a certaintie of double Winchester: but though they raysed the lower, they cannot abate the higher to this proportion: and yet from the want of this reformation, there ensue many inconueniences: for the Farmer that hath the greatest bushell at the market, maketh a price for the lesser to follow with little, (or at least) no rateable deduction. Besides, they sell at home to their neighbours, the

P 2 rest

Figure 3.1 Richard Carew, *Survey of Cornwall* (London 1602), pp. 53d–4 from a microfilm copy in Exeter University Library.

'commotions'.[5] He also found time and energy in an incredibly busy life to rescue and put in order the city's archives, for which alone historians must be grateful, but furthermore he used them to compile his still unprinted but extensively-quarried 'Annals' of the city.[6] In the pages of this wonderful compilation every Tudor mayor of Exeter, and some even earlier, is known to us as a man of flesh and blood, with nothing hidden of their strengths and human frailties, at least as they appeared to the remorseless Hooker, to whom all men were either 'godly' or 'Romish'. It is in his 'Annals' that his dry wit is most in evidence. He claimed to have been denied access to the records of the Exeter diocese, but nevertheless wrote and published in 1578 *The Lives of the Bishops of Exeter* which still finds a place on the shelves of twentieth-century historians.[7] He also found time to sit not only in several English parliaments (which were mercifully of short duration) as representative of Exeter, but also in what seemed to him an incredible Tower of Babel, the Irish Parliament in Dublin, the opportunity occurring while he was there in the service of Sir Peter Carew (of the Devonshire family).

Besides his work on the history of Exeter and of the diocese, Hooker also turned his attention to a description, set against an historical background, of his native county. His somewhat grandly-styled 'Synopsis Corographical of the county of Devon' was, in many ways, his most original work, and probably the most laborious. It was a long time on the stocks, probably begun before Carew's *Survey of Cornwall*, to which it serves to some extent as a companion-piece. It was revised many times and only reached the state in which it has come down to us shortly before his death. He may, even then, have contemplated further revision. Two almost identical manuscript copies survive. One is in the British Library and carries the date 1599/1600. A very small part of this was published in 1915.[8] The other, slightly longer manuscript copy was, in 1686, in the library of the Revd John Prince of Berry Pomeroy; after various travels, however, it is now safely back in the county archives (Figure 3.2).[9] In fact after Hooker's death, as was usual with such manuscripts, both copies went into private circulation. The shorter version was read and highly commended by Judge Doddridge after having been offered on the old man's instructions to a 'person of honourable place in the commonwealth', possibly Sir Walter

Raleigh, Hooker's long-time friend and former patron, no doubt with a view to promoting its publication. Why Raleigh failed the man he had once employed to research his own pedigree, although at about the same time accepting the dedication of Carew's *Survey of Cornwall*, will probably never be known. Incidentally Carew and Hooker were acquainted and expressed mutual admiration, but although Hooker was well-known in London publishing circles, he was never, like Carew, a member of the Society of Antiquaries. One feels that, given the chance, he would, for all his busy life, somehow have made the time! He might also have been a great nuisance! He had a high opinion of himself and brevity was not his natural wont.

Was the 'Synopsis' in fact publishable in the state in which Hooker left it? Whereas Carew's *Survey of Cornwall* was a polished work of literary merit such as any publisher would seize upon, Hooker's 'Synopsis' is very largely a work of reference. It opens with a disappointingly brief though brilliant general survey of the county's human and material resources, but with less comment than is to be found in Carew. There then follows a series of appendices, invaluable in themselves but factual rather than analytical, more like a commonplace book or scholar's notes than a literary text. Not that Hooker isn't, most of the time, a master of clarity, and in places of literary style. It depends, of course, on what the reader wants. Hooker can tell you how many markets there were in Devon. He knew what each parish paid in tithe wool, information he recognised as an invaluable historical source, but stopped short of divulging details lest he offend the recipients. In other respects he felt no such inhibitions, his listing of parochial assessments to the parliamentary subsidy showing the bureaucratic mind at its most informative. There is indeed plenty of evidence of Hooker's appreciation of archival evidence, to the extent that in among his listing of the county's chief officers he inserts full transcripts of letters patent of appointment, indicating, incidentally, that while in London he must have fitted in visits to the office of the Master of the Rolls. There is no evidence that he ever saw any of the county's own records, these probably being out even of his reach in the private residence of the clerk of the peace.

But not all Hooker's appendices are dry as dust. Interspersed among the reference matter are many vivid descriptive passages, for instance

Cornewall and so theres ptes it is inclosed wth the seas
and waters: onely the East pte lyeth vpon vpon the
meane Landes and borders of Durotrigia and Belgia and
theise marches beinge also full of Vales, hilles, Rockes, &
stones, is very easylie to be made stronge and fortified
agaynst the invasion of any enemye: yf the disloyaltie
of the subiectes do not cause the contrarye./. The
whole province and countrie wthin theise boundes, is in
greatnes, the seconde to the greatest, in this Lande &
is alltogether or for the most pte, wilde full of wastes
hethes and Mores, uphill and downehill amonge the
Rockes and stones and the [illegible] longe craggye and
very paynfull for man or horse to travell. as wth
all straungers travellinge the same, can wytnes it /
ffor be they never so well monted vpon their fyne
and deyntie horses, out of other countries, after that
they have travelled in this Countrie but one Iourney
they can forbeare the seconde./. And therefore so
much the lesse passable for the enemye wth his
troopes and impedimentes of warres X It was
in tymes past all forest and wood Land countrie even
as the now Citie of Exceter wch is the metropole
of the same can wittnesse it wch was called caer
penhuelgoyte that is to saye the cheife Citie in the woode

Figure 3.2 Page from one of two manuscripts of John Hooker's 'Synopsis Corographical of Devonshire', *c.*1599, Devon Record Office Z19/18/19 fo. 9.

In this passage (lines 7–20) Hooker describes the size and hilly character of Devon and the problems faced by travellers:

> *The whole province and countrie within theise boundes is in greatnes the seconde to the greatest in this lande and is alltogether, or for the most parte, wilde, full of wastes, hethes and mores, uphill and downhill amonge the rockes and stones, and the pennettes [?pavements] longe, craggye and very paynfull for man or horse to travell, as which all straungers travellinge the same cann wytnes it, for be they never so well monted upon theire fyne and deyntie horses out of other countries, after that they have travelled in this countrie but one journey, they can forebeare the seconde. And therefore so mich the lesse passable for the enemye with his troopes and impedimentes of warres.*

brief biographies of Devon's literary men down the ages, concluding with very precise details for the edification of posterity of his own substantial achievements in the field. Most of the county's larger towns are described, the contents being of varying interest. One which he apparently knew well, perhaps through having visited it frequently in the course of yet another of his duties, that of a judge of the court of Raleigh's vice-admiralty, was Barnstaple in north Devon which he described as:

> a verye ancient towne . . . and maye be equall with some cities for it is the cheffe emporium of that countrie and most inhabited with merchantes whose cheffest trade in tyme of peace was with Spayne . . . It is a clene and sweete towne, very well paved . . .[10]

Where he could not improve on what he had read he copied without acknowledgment. Many quite recent social historians have quoted his categorisation of local society into four degrees (noblemen and gentlemen; merchants; yeomen; and labourers) without, apparently, having noticed that it was lifted almost verbatim from the Essex parson, William Harrison's *Description of England*, with which Hooker was very familiar.[11] But Hooker was more often sinned against than sinning. Perhaps four hundred years later, rather than be censorious we should be making some amends by at long last publishing the 'Synopsis' as Hooker left it, establishing thereby his priority as the father of Devon's history. As Carew said of his *Survey of Cornwall*, the necessary preparation for the press is a task 'long since begun [and] a great while discontinued'.

Thomas Westcote

One of those who followed Hooker's lead was Thomas Westcote, who completed the manuscript of his own 'Survey of Devon' in about 1630. He complained, very peevishly, that he had been denied access to records in Exeter used by Hooker, who, he declared, had only 'chalked the way'. He even claimed that he had been unable to consult Hooker's text, which he clearly regarded as his right, although the evidence that he did indeed gain access is laid bare on page after page. For example he described Barnstaple as:

> a very ancient borough . . . It is one of the eyes of the country, and the northern emporium, and may, without offence, be compared with some cities . . . The inhabitants trade into foreign countries, especially in regard of the [international] situation, to Spain and to the Islands. The streets are somewhat low, yet well paved and thereby clean and sweet in all weathers.[12]

Compliments such as this are, of course, worth repeating, 'imitation' being 'the sincerest form of flattery'. But it really was going a bit far to follow Hooker in characterising the tinners as chaps who drank from their spades, to say nothing of his division of contemporary society into gentlemen, yeomen, merchants and husbandmen. But Westcote, who incidentally was a minor landowner in Shobrooke, north west of Exeter and took to his study after having fought with Drake on the Portuguese expedition of 1589, was the better man in some respects, his style being not that of the trained lawyer or even the scholar but that of the poet. Listen, not least for its anthropomorphism, to a sentence from his description of one of the tributaries of the river Exe above Newton St Cyres: 'Creedy here seems to vaunt of the fruitful soil he passeth through, which never proves ungrateful to the labourer for his pains nor deceiveth the husbandman's hope of . . . increase'.[13] What better way of indicating how the rich water-meadows made, and still make, the lower Creedy valley such valuable farmland, a lesson, incidentally, which Elizabethan farmers elsewhere in England were only just beginning to learn.

Westcote was also more sensitive than Hooker to the consuming interest of his potential readers, not only listing the gentry but including pedigrees of no less than three hundred of them. Not all of these were accurate of course, but the early Stuart gentry did not regard themselves as being on oath where their ancestry was concerned. Westcote gives some indication of his methods of obtaining information when he writes of 'meeting casually or purposely with some of the heirs of such families . . . and by way of discourse or some entreaty . . . I have moved them to give me sight of their pedigree'.[14] He also tells of one gentleman who, to the amusement of others present, declared that he did not know his father, who died when he was young, and, adds the mischievous old soldier, 'indeed he spake no more than the world knew to be true'. On the other hand he admitted somewhat diffidently to searching epitaphs, none of which, he remarks, are 'so silly or simple that somewhat may be learned out of them, or some while spent in laughter at them'.[15]

Like most of his contemporary topographical writers, Westcote had little grasp of history, but he did know about Henry VIII's, or rather Thomas Cromwell's, short-lived Council of the West, adding the shrewd comment that Devonians, if not Cornishmen, had little stomach for home rule.[16] But, like so many of his fellow gentry, and with much reference to classical authors, he was totally convinced of the superiority of Devon to any other English county, and in describing its resources, closely echoed Hooker's words, that 'this litle corner of this land can lyve better of itselfe without the rest of this land than all the resydue can live without it'.[17] Carew too was proud of his county but no doubt thought the superiority of the Cornish to be self-evident.

But one last word on Westcote. Like most Elizabethan country gentlemen, he wrote primarily for his own amusement and that of his friends and did not aspire to publication. Not until over two centuries later was one of the many copies still in circulation prepared for publication, fortunately by that excellent Victorian scholar Dr George Oliver, who made no secret of the fact that he had corrected such errors as he detected and extended some of the pedigrees. But, less forgivable, he and his collaborators rendered the text into modern spelling.[18] Someone should seek out one of the surviving copies of Westcote's original text and give it to us in print as he left it.

Tristram Risdon

Another writer in this field was Westcote's slightly younger contemporary, Tristram Risdon of Winscott near Torrington (*c.*1580–1640). His 'Chorographical Description or Survey of the county of Devon . . . [compiled] for the love of his Country and Countrymen' also circulated in manuscript for nearly a century until it was shamefully hacked about by a rogue publisher in the early eighteenth century and only finally tidied up and amended in the form now most readily available nearly a hundred years later.[19] Only when all the extant versions, in print and in manuscript, have been collated will it be known what Risdon himself actually wrote. Again his general description has many echoes of Hooker and the three hundred pages of topographical detail which follow make extremely tedious reading, unredeemed by Westcote's style.

William Pole

Consideration of Risdon does, however, take us back to where we began, for he refers gratefully to the compilations of Sir William Pole, 'from whose Lamp I have received Light in these my Labours'. With Pole, he of the jolly face at Antony, we are coming nearer even than with Hooker to one of the proper preoccupations of modern professional historians, the building up of a body of original documents. Being a man of his time, the material was largely concerned with the genealogy and landed possessions of Devon's aristocracy and gentry, and he found no place for the rest of society. His collections, including original manuscripts, were made very largely before the Civil War and some, perhaps a great many, were lost during that conflict. What remained eventually accompanied the family deeds to Antony, most of them having already been published under a title more modest than those of his fellow antiquaries.[20] Already widely consulted, their appearance in print was to prove invaluable to nineteenth-century topographical writers, including of course the Lysons brothers. To each and every one of these early scholars it is right that members of the Centre for South-Western Historical Studies should pay homage. They did far more than merely 'chalk the way'.[21]

4

Somerset Topographical Writing, 1600–1900

ROBERT DUNNING

'Many attempts have been made towards a description of Somersetshire', wrote Richard Gough in 1780, 'but without success'.[1] Gough, writing a general survey of works on British topography, naturally concentrated his attention on published material, and to that extent his assertion was correct. Somerset had not been ignored by topographers and antiquarians; rather, either by temperament, circumstance, or design, they had not appeared in print. Yet the work of several unpublished Somerset antiquaries was known and valued in their own time, and much of it still survives, in range quite remarkable and in content high in quality and sometimes unique in survival.

Among the successors of William Camden, John Stow and Richard Carew who met in the house of Sir Robert Cotton in the 1580s and called themselves the College or Society of Antiquaries was Thomas Gerard (1582–1634) of Trent (Dorset, formerly Somerset). He inherited a modest estate as a child, was educated at Oxford, and married the daughter of a Dorset squire. His authorship of surveys of Somerset and Dorset was not established until 1897[2] but the survival of a unique copy of the Somerset survey, known as the 'Particular Description' and bound up with several surveys by John Norden,[3] is of some significance, for in the forty years between Norden and Gerard a radical change in concept had taken place. It was a change made possible both by a growing corpus of materials and by a widening consciousness brought by the cartographer and the professional genealogist. Gerard openly acknowledged information from a variety of medieval chronicles in editions produced by his contemporaries; he cited escheators' rolls and inquisitions post mortem, patent, close and plea rolls, Domesday Book and the Nomina Villarum from the public records, cartularies and charters in private hands; and the works of John Leland, Lambarde, Camden and Selden.

Gerard's 'Particular Description' in its surviving form covers only one half of Somerset, although internal evidence proves that the whole was completed. It bears the date 1633. 'Particular Description' is an apt title, for it is a journey through the county following streams and rivers, with family history as the theme. Here is the topographer and antiquary combined, an author whose eye for land and its produce is as keen as his recording of genealogy and blazon.

> Seated [Gerard says of the village of Raddington] in somewhat barren Countrey whence it seemes it tooke name for Radin and Redin which the Britanes signifies Fearne, and that ever shewes a poore ground. This place had Lords of the same name of which John de Radington lived in Edward the 2 tyme, but more of them I cannot say. Having thus lead you through this angle of the county and followed the river Ex, till it forsooke mee and entred into Devon; let me now intreat your company backe unto the North shoare.[4]

Under Porlock on that north shore Gerard found a park with a chapel dedicated to St Culbone, 'a Saint I assure you I am not well acquainted withall, and therefore can say no more'. But he was soon able to overcome his modesty as he noted the coats of arms of Rogus, Cheseldon, Wadham, Blewett, Mohun and Rogers, all families connected with Porlock manor.[5] Gerard's entry on Langport quotes Leland's 'Commentaries' on Henry I's proposal to remove the Benedictine community there from Muchelney as part of his penance for the murder of Becket and also part of a deed of the late twelfth century then owned by Sir Thomas Lyte. But he also knew the surrounding countryside and recorded how the winter Saturday market in the town was full of fowl and eels. The eels he could not commend, but . . . 'marry the fowle is fetched hence farr and neere, but the waters being abroad such as are sent for it many times missing the Cawsway goe a fishing instead of getting fowle'.[6]

Thomas Gerard died in 1634, within a year of the completion of his work,[7] four years before those seminal meetings between William Dugdale, Roger Dodsworth and Sir Henry Spelman which had such 'profound results upon the later development of English medieval scholarship'.[8] The political and religious upheavals of the next few years provided a background contrasting markedly with the settled social scene of privilege and heredity which Gerard unconsciously described through feudal descent and heraldic escutcheon. He would have been proud that his house at Trent sheltered the fugitive Charles II in 1651 after the battle of Worcester[9] but disturbed at the insecurity of the Crown. Yet the years of Revolution in the 1650s and the political controversies which continued for the next seventy years gave rise to a great era of English scholarship far in advance of the topographical surveys of the past; scholarship fed on the profundities of Dugdale and his circle.

The Somerset-based scholars of the years immediately after the Restoration, men such as Joseph Glanvill (1636–80),[10] John Beale (*c.*1613 – *c.*1682)[11] and John Beaumont (d. 1731) of Ston Easton, seem to have been principally interested in natural history and witchcraft, although Glanvill contributed 'notices' on Bath to the *Philosophical Transactions of the Royal Society*, of which all three were fellows.[12]

A fourth scholar, the Somerset-born William Musgrave, could hardly have included Glanvill and Beale when he wrote in 1684 from Oxford of a 'design' recently reported to the Philosophical Society there which was being 'carried on by several of the most learned men in Somersetshire to write the natural, civil, and ecclesiastical history of the county'.[13] John Beaumont certainly submitted a draft proposal for a natural history of the county to the Royal Society at about that time, having already published material on Wookey Hole and other Mendip caverns.[14] The moving spirit of the larger enterprise, the 'principal undertaker' who was also to 'digest' the work of its contributors, was Andrew Paschall (d. 1696),[15] a fellow of the Queen's College, Oxford from 1653 and rector of Chedzoy near Bridgwater from 1662.[16] Paschall was one of Aubrey's correspondents[17] and his surviving papers reveal interests including Wilkins's Universal Language, drainage of the

Somerset Levels,[18] a Roman pavement at Bawdrip,[19] gold mines at Glastonbury, medieval remains on the site of Athelney Abbey, and natural phenomena including thunder on the Poldens and Siamese twins born at Isle Brewers.[20]

The thunder and the 'monstrous birth' afforded the opportunity for the loyal Anglican Paschall to suggest that they in some way presaged the disaster known to historians as Monmouth's Rebellion. He used them to introduce his longer history of the event which culminated in the battle of Sedgemoor in which his parish was intimately concerned, and was evidently based on first-hand accounts given to him by his parishioners.[21] Both accounts and an accompanying map bear evidence of intimate local knowledge. After his death in 1696 Paschall's books and papers are said to have passed to Dr Roger Mander, Master of Balliol and a Somerset native, though Thomas Coney, his successor at Chedzoy, had some fragments of his library.[22] Some of his papers were published by Richard Rawlinson.[23]

The political controversies which gave rise to Monmouth's Rebellion were enflamed rather than calmed by the 1689 Revolution, and out of that particular event more than one scholar emerged. Among those who could not accept the new order was George Harbin (d. 1744), an Essex man but of good Somerset and Dorset stock, who after Cambridge became chaplain to Francis Turner, bishop of Ely, one of the Seven Bishops.[24] With Turner he found himself unable to take the Oath of Allegiance to William and Mary, and for several years led the rather shadowy and wandering existence of most Nonjurors. In 1699, on the recommendation of Thomas Ken, he became chaplain to the first Lord Weymouth at Longleat. Ken thought him 'an excellent scholar . . . of a brisk and cheerful temper' and surviving correspondence suggests that for a year or two Harbin was interesting himself in the history of the Reformation. From about 1703 he ceased to be chaplain but continued at Longleat as librarian and became a leading protagonist of the Stuart cause. In 1713 he published, anonymously, a highly influential tract entitled *The Hereditary Right of the Crown of England asserted, the History of the Succession since the Conquest Cleared*, which was a thinly veiled statement that the Act of Settlement was invalid.

The research entailed in this work was based in part on books in the library at Longleat, but also relied on manuscripts 'in the possession of

Lord Treasurer Harley', some indication of the political circles in which Harbin moved. The death of Queen Anne and the succession of the Elector of Hanover put paid to any hopes that Harbin and his Nonjuring friends had of preferment. He himself seems to have been librarian to Lord Lansdowne for a short time but his letters show him visiting his own or Thynne family relations; a visit to the Isle of Wight produced genealogies of local families and notes on local topography. At other times he was copying out Somerset visitation pedigrees and other manuscripts from the Harley collection and corresponding with Harley cronies like John Anstis, Garter King-at-Arms.

Thomas Hearne was not entirely sure when he first met Harbin that he was a real scholar. He was certainly a 'useful assistant' to men like Atterbury, providing them with 'choice papers' formerly belonging to John Selden which he had found in Sir Matthew Hale's collections.[25] The two were in frequent correspondence over the purchase of texts from Hearne including copies of Camden's *Britannia* for himself, Bishop Hooper of Bath and Wells and Lord Foley.[26] But still Hearne had doubts about Harbin's scholarship: he could not believe him capable of writing *Hereditary Right* and concluded that he only edited it.[27]

Harbin's social and political contacts gave him access to private libraries as well as public records. He had valuable papers on the life of Mary, Queen of Scots, a copy of *Holinshed's Chronicles* which Lord Weymouth had borrowed from the earl of Cholmondley, a copy of a record he saw at the Tower, and copies of most of the important charters in the cartulary of Glastonbury Abbey at Longleat.[28] Harbin duly sent his Glastonbury material to Hearne in March 1723, but a month later, on its return, part was missing, much to Hearne's distress since he had had no time to transcribe that very portion.[29] It seems to have been some years before the correspondence between the two men was renewed and only in 1730 because Hearne wanted to borrow another transcript. He might even have been contemplating some sort of memoir for he asked a Cambridge correspondent to enquire when Harbin took his degree and described him as 'a right worthy, learned, honest man . . . well versed in heraldry and antiquity'.[30] He hardly needed to be reminded later in that year by John Anstis that Harbin was a 'gentleman of so communicative temper . . . that he will readily impart all the notices he hath to you'.[31]

Harbin's collections are scattered, partly thanks to human failure, partly by the fact that he left no direct descendants. Papers on Somerset families, some of them 'drawn up by Thomas Carew', were bought from the Phillipps collection by Prebendary Bates Harbin,[32] and include extracts from the MSS of Richard Rawlinson, John Anstis and Sir William Pole, and Harbin's own extracts from the Harley collection, Camden and Dugdale. Somerset gentry frequently visited Harbin at his home in King Street, Westminster, conveniently placed for access to Lord Oxford's library where he spent so much time.[33] Among them were Thomas Carew of Crowcombe and Thomas Palmer of Fairfield. Carew was probably his most intimate friend, regularly writing to him about political matters[34] and Harbin's manuscripts include papers 'drawn up by Thomas Carew and given me'.[35] John Bampfylde, another of the group and husband of Harbin's niece, wrote in March 1737 to John Strachey of Sutton Court, then actively engaged on a projected History of Somerset, suggesting that Strachey should consult Harbin, who had many manuscripts on county families, copies of which he had formerly lent to Mr Palmer.[36]

This constant lending of papers has led inevitably to loss, but in at least one case to fortunate survival. The cartulary of Athelney Abbey has not been traced since it was in the possession of Sir William Wyndham at Orchard Wyndham in 1735. Harbin made a transcript of it, part of which passed from the Malet family into the Phillipps collection and is now in private hands. The elusive first thirty-two folios in Harbin's own hand, which include two charters of King Alfred, have now been traced to the collections of Harbin's friend Thomas Carew.[37] Carew's papers remained at the family home in Crowcombe and did not suffer such dispersal.[38] They include much original material as well as copies made either by Carew himself or by others for him, and they deserve further careful study. Among them are papers about parliamentary elections at Milborne Port between 1679 (when Carew's great-grandfather was sheriff) and 1741.[39]

It is not clear whether Carew contemplated more than a collection of material, but in his comparative youth Thomas Palmer, his fellow visitor at King Street, had certainly intended to do so. Thomas Hearne dined with the 32-year-old Palmer in January 1717 and thought him 'a very good scholar . . . a good antiquary . . . and an excellent herald'.

Palmer himself was modest enough to speak of a county history 'now doing by several hands', although it was clear that he was the principal instigator and had the 'collection' in his possession.[40] The history was reported to be almost finished in 1718 and was considered in prospect 'an excellent work and of great use and entertainment to the public'.[41] The report proved premature. By 1732 it was admitted that the work 'had been laid aside' but Palmer still continued to collect material, usually from the Cotton library.[42] Indeed, a year later and only a few months before Palmer's death, it was sadly admitted that despite his great collection of manuscripts, the history had 'not got above 5 miles round Glastonbury', and its author had 'been weary several years' and had locked his papers up 'so that we must not look for anything from him'.[43]

After Palmer's early death, his books, and perhaps many of his manuscripts, were bought by Sir William Wyndham and were added to the library at Petworth.[44] Some manuscripts of the history remained at Fairfield among the estate records; carefully compiled drafts in different stages of completion, principally concerned with the hundred of Williton in which the family holdings were most extensive.[45] The drafts are based on Palmer's own family archives, on the great collection of Sir William Pole of Shute, and on original charters in his neighbours' strong-rooms, such as a manuscript owned by W. Simmonds of Bridgwater which he saw in 1712 or papers of the St Albyn family of Alfoxton.

Most of Thomas Palmer's work on Somerset county history seems to have been done before 1720 and it is improbable that it was unknown to others working in the same field. In 1721 Thomas Hearne referred to a group of men who were 'collecting what remains they can of antiquity relating to Somersetshire, not with a design to publish any work, but to have them lodged in some safe place where they may be preserved and consulted upon occasion'. This work, the suggestion of Dr Hooper, bishop of Bath and Wells, was being undertaken by Thomas Ford and others.[46] Ford, vicar of Banwell from 1713 until his death in 1746 and a canon of Wells from 1721,[47] had himself earlier worked on lists of the deans and prebendaries of Bristol in connection with Le Neve's *Fasti*.[48] Hearne considered Ford 'a tolerable good drawer' and linked him with Richard Haynes of Bristol, whom Ford thought 'an excellent scholar' and whom Hearne said had also been collecting Somerset material.[49] Ford wrote a preface to John Whitson's *A Pious Meditation . . .*, published in Bristol in 1729 from a manuscript in Haynes' collection.[50]

The early 1730s saw the compilation, probably by an official hand, of a financial and statistical survey of the county, principally recording how each parish and hundred was rated for gaol, marshalsea or hospital money and for county stock.[51] The author includes occasional and unfortunate essays into history and etymology, but the contemporary information on sheriffs' tourns, quarter sessions and hundred courts is valuable and often unique, and there are lists of county bridges and notes on commissions of sewers and justices' divisions.

Another scholar whose work came to fruition in the 1730s was John Strachey,[52] who was only a few years younger than George Harbin and who died a year before him in 1743. Strachey's father had been a close friend of John Locke and he himself was heir to the considerable Sutton Court estate which he left heavily burdened with debt. His interests were not, like Harbin's, politically motivated; he was a landowner, in some ways typical of his class, remembered for his 'affability and condescension' and his 'many civilities'. And yet he was also untypical, for he combined a natural interest in the genealogy of gentle families and the history of their estates with a fascination for the landscape around him. He may reasonably be claimed as 'one of the earliest English geologists who was an observer and recorder of geological facts, rather than a theorist. He made what were, for the time, outstanding contributions to stratigraphical geology'.[53] For those contributions he was elected a Fellow of the Royal Society in 1719.

Rather late in life, when he was about sixty, Strachey turned seriously to history. He was, in fact, rather archivist than historian, and his knowledge of the public records is as impressive as his scientific expertise. Thomas Hearne, who first met him in 1730, described him as a gentleman who delighted in antiquities,[54] and thought his work on coal mines showed him to be 'an ingenious man and of good skill in affairs of this kind'.[55] 'A good sort of man', declared Palmer, 'notwithstanding a whig.'[56]

In 1731 Strachey produced *An Alphabetical List of the Religious Houses of Somerset*, which was published by Hearne in his edition of

Hemingford's Chronicle;[57] but far more important was *An Index to the Records, with Directories to the Several Places Where they Are To Be Found*, which appeared in 1739.[58] It is a kind of 'historian's encyclopaedia' listing topics from abbeys to writs and it included, incidentally, a glossary of Latin surnames and English place-names 'as they are written in our Old Records explained by Modern Names'.

The working papers for this volume show what a massive undertaking it was, for it involved both a knowledge of the physical difficulties facing the would-be searcher as well as some understanding of the character of the records themselves. There is a letter,[59] for instance, to 'R. T. Esq.' 'concerning the condition of certain Common plea rolls, King's Court rolls, etc. dating from Ric[hard] I' which Strachey found in Tally Court 'in a much better condition than the rotten Essoins in your own office at the White Tower'. Elsewhere there is a note concerning the incorrect labelling of the plea rolls before the Great Charter of 9 Henry III and a chart showing the arrangements of labels on press or shelf accommodation in the volumes of the Entries of Proceedings in the Court of Wards and Liveries.[60] His 'account' of the records in the Receipt of the Exchequer, which then included those of Star Chamber, Domesday Book, and the rolls of both King's Bench and Common pleas from the Conquest to the reign of James II,[61] indicate a remarkable grasp of the sources of English history, both local and national. And these notes and accounts were accompanied by extracts of Somerset entries from the plea rolls, rolls of parliament, patent, close and Curia Regis rolls.[62]

All this activity may have run in parallel with his scientific studies but surviving letters suggest that the serious work on his 'collections' had begun by 1731 and that in 1732 his map was 'going to be printed'.[63] In the event its publication was postponed for, as Palmer told Hearne, Strachey had recently gone abroad 'by reason of his perplexed affairs, he having about 14 children living and none good'.[64] The map, the first large-scale map of Somerset, eventually appeared in 1736.[65] In the same year Strachey issued proposals for a work to be entitled 'Somersetshire Illustrated'.[66] The proposals appeared in July 1736 and in August Thomas Carew wrote to him, offering to lend his papers.[67] Thomas Palmer had already, two years and more before, given a brief description of his own progress and suggested various antiquities which ought to be noted on the map.[68] The result, after the archival roar, was a pathetic whimper: sixteen sections of octavo manuscript and rather less in folio.[69] Perhaps Strachey was the victim of his own exacting standards and was unable to take the work of others on trust. That, indeed, was the purport of his introductory remarks to his *Index*:

> Many Records of the same Nature which ought to be collected into one place, are scattered in several Offices . . . Whoever has had occasion to search for these valuable Treasures, must be sensible of the great Trouble and Difficulty of finding them . . . These Difficulties have occasioned many Lawyers and Historians to commit great Mistakes, by taking things on the Credit of those who wrote them, instead of having Recourse to the Originals themselves.

In 1742 the *Sherborne Mercury*, then the leading newspaper in the county,[70] thought to steal the thunder of antiquarians like Strachey by offering its gullible readers 'A compleat history of Somerset Shire' to be issued in fifty-two weekly numbers. The contrast between this and the work of Strachey and his friends could not have been more marked for it was, in fact, a re-issue, with some omissions, of the Somerset part of Thomas Cox's *Magna Britannia*, which had appeared in 1727.[71] The only new material was some up-to-date information about the county's charity schools.

But the work of earlier antiquarians was not forgotten, and when in 1781 John Collinson first directed his thoughts to a Somerset History it was to the papers of John Strachey that he first turned.[72] Access to them was refused and Collinson continued instead with his other projects. He was a most industrious man who while still an undergraduate at Oxford had begun collecting material for a history of his native Wiltshire. He evidently had no time to take a degree, for in 1779 he both published a book entitled *The Beauties of British Antiquity: selected from the writings of esteemed antiquaries with notes and observations*, and successfully studied for deacon's orders.

After a short curacy at Marlborough he moved to Cirencester where he married the daughter of a prosperous newspaper proprietor, took on the very light duties of curate in the parish, and spent most of his time in Earl Bathurst's library at Cirencester Park. He may well have acted as domestic chaplain to the family, and it was almost like Harbin at Longleat over again. It was at Cirencester that the idea of a Somerset History first arose; and there, sharing the society of Samuel Rudder, Gloucestershire's historian, and Joseph Kilner, another like-minded clergyman, Collinson pursued his local studies assiduously enough to be recommended to the Society of Antiquaries of London by Daines Barrington, William Boys and Edward Hasted. Collinson was still living in Cirencester (though drawing an income from the vicarage of Clanfield in Oxfordshire) when in 1784 he again issued Proposals for a History of Somerset, soliciting subscriptions for a work which would combine 'authentic records' with an 'actual survey'. The historical and ecclesiastical parts were to be his own work, the topography and natural history were to be contributed by Edmund Rack, Secretary of the Agricultural and Philosophical Societies of Bath. The work was already 'in considerable forwardness' and Collinson claimed to have received 'considerable assistance from divers gentlemen'; and as an earnest of things to come specimen histories of Chilcompton and Porlock appeared.[73]

Hitherto the individual contributions of Collinson and Rack could not be separately assessed, for after Rack's death in 1787 his work was absorbed with only general acknowledgement. But in 1994 his detailed survey of the county came to light, partly in private hands but largely among the Smythe family papers formerly at Ashton Court near Bristol.[74] Preliminary study[75] indicates that Collinson rejected good detailed descriptions of church interiors, and omitted informed comments on land use, sizes of settlement and other material.

During the later stages of work on his history, Collinson removed from Cirencester to become vicar of Long Ashton and perpetual curate of Whitchurch in 1788. The work was completed at the beginning of 1791 but was not published in its entirety until the following year. Collinson himself died of a 'lingering illness' eighteen months later at the age of 36, a man 'whom to know was to respect and love'.

The reviewers certainly did not respect him, and for two years every number of the *Gentleman's Magazine* included several statements for the prosecution and only a few for the defence. Many of the writers were 'critical nibblers', others had justice on their side. A century later Sir Henry Maxwell Lyte probably came nearest to the truth when he wrote that the work was 'meagre and in some sections fundamentally incorrect';[76] but Maxwell Lyte had worked on parishes for which the Luttrell manuscripts survived in great quantity. For other places Collinson looks much less meagre, and his preface and footnotes, such as they are, are the perfect answer to one critic who accused him of being 'a perfect stranger' among the public records at the Tower and Rolls Chapel and at the British Museum. And Collinson's answer would have been that Hugh and John Acland of Fairfield had allowed him to use Thomas Palmer's manuscripts, that James Bernard of Crowcombe had lent him two volumes of Thomas Carew's papers, that the marquess of Bath had opened the library at Longleat for the perusal of the Glastonbury cartulary and the Red Book of Bath, and that the bishop of Bath and Wells had made him free of the Diocesan Registry at Wells. Footnotes citing 'original deeds owned by Henry Hippisley Coxe', 'original papers in the possession of Sir J. H. Smyth', 'a deed in the church chest at Dundry', Cole's Escheats in the British Museum, ancient charters in the Cotton Library, Rawlinson's Book of Inquisitions in the Harleian Library, or the will register Horne in the Prerogative Court of Canterbury are not sources of an incompetent antiquary confined to printed books. They may, of course, actually be cited at second or even third hand. Despite the criticism levelled against the work, Collinson has had his ardent supporters and John Rutter, a Shaftesbury publisher, claimed to have been inspired by him and Sir Richard Colt Hoare's *Modern Wiltshire* to produce in 1829 a survey of parishes around Weston-super-Mare.[77]

Colt Hoare was also the inspiration for a new county history proposed on a large scale by the Revd William Phelps, another clergyman whose parochial duties lay rather lightly upon him and who was already the author of a volume on botany.[78] In 1835 Phelps announced and almost immediately began to publish a county history which was to comprise both general chapters and parochial histories. The enterprise presumably foundered for financial reasons soon after 1839,[79] but meanwhile eight parts appeared comprising the general chapters and the histories of parishes in thirteen hundreds in the centre and south-east of the county.[80] The general chapters, it should be said, may safely be ignored, but the topographical surveys, based in part on the answers to questionnaires,[81] are a significant advance on Collinson.

Calls for a new county history were made by Dr William Buckland, dean of Westminster and distinguished geologist, when he addressed the newly-founded Somerset Archaeological Society in 1849, though his proposals were quite modest: 'a small monograph' to include 'subterranean antiquities' on the one hand and 'present natural history' on the other, the whole work under a 'properly qualified person' who would organise the returns to questionnaires sent in the first instance to all clergy and magistrates and later to any other interested resident. A year later the survey had already proved a failure.[82]

Ten years on and the members of the archaeological society came to the conclusion that the best way forward was to support their library and museum, publish papers and set up a committee to revise Collinson.[83] Ralph Neville Grenville, president of the society in 1860, 'trusted that as Hutchins's *Dorsetshire* was being brought out in an improved and valuable form, so Collinson's *Somersetshire* would be brought out in a greatly improved and much more valuable form'.[84] At that same annual general meeting, however, the rather more practical general secretary was forced to admit that 'no great advance has been made towards the attainment of our chief desideratum—a good County History', particularly because of financial difficulties caused by the defalcations of the late curator who had made off with members' subscriptions.[85]

For almost a century, people clung to Collinson, many urging simply a better index rather than revision, and the archaeological society, with broken promises on its conscience, produced an index in 1898.[86] In that same year Henry Hobhouse returned to the theme of his presidential address to the society in 1890, the need for a good county history; and in 1899 Sir Edward Fry called members' attention to work then proceeding on a large-scale history of Northumberland.[87] No-one seemed to know, or at least the society's secretary did not see fit to record, that another enterprise was also in hand.

How and when the *Victoria County History (VCH)* was first heard of in Somerset will never be known, though Herbert Arthur Doubleday and Laurence Gomme probably issued their general prospectus *after* Sir Edward Fry had delivered his presidential address in 1899.[88] The curator of Dorset County Museum was clearly surprised to read that Dorset's History was in 'in active preparation' and before the end of the year wrote to ask if Doubleday 'would be so obliging' as to inform him if it would soon be published.[89] Certainly in the autumn of 1900 Doubleday wrote not to the Somerset Archaeological Society but to county society. The earl of Cork, Lord Lieutenant, had agreed to act as chairman of the county committee 'with a view to gaining access to all the materials necessary for the compilation' of a History of Somerset; and Doubleday then set about recruiting others. The bishop of Bath and Wells expressed himself 'very glad' to do all he could to assist and regretted that he had overlooked Doubleday's letter 'through my having placed it among circulars, probably owing to the printed matter at the top'. The dean of Wells returned the original letter with a note offering his help but only if 'no monetary obligation of any kind will be incurred'. Sir John Horner of Mells preferred a revision of Collinson and Henry Hobhouse was cautious, telling Doubleday that he would not commit himself before he knew 'something of your scheme and your editor'.[90]

So a committee[91] of Somerset's great and good was brought together, and it included (such is the persistence of Somerset families) Sir Edward Strachey, a direct descendant of the topographer John, and Sir Thomas Dyke Acland, a distant relative of Thomas Palmer. Another member of the committee was the Revd E. H. Bates who, Henry Hobhouse would have been delighted to discover, was editor of the Somerset History in all but name, Doubleday's local contact and contributor of the texts of Domesday and the Geld Inquest. Bates's qualifications were impeccable. He had been an active member of both the Somerset Archaeological Society and of the Somerset Record Society since 1886 and it was he who in 1899 had edited for the Somerset Record Society all that was then known of George Harbin's transcript of the Athelney cartulary,[92] and in 1900 the pioneer work of Thomas Gerard.[93] The first work was entirely appropriate, for his second name was Harbin and in 1909 he took it as an additional surname when he inherited from his uncle the ancestral estate of Newton Surmaville near Yeovil.[94]

The *Victoria County Histories*, planned like Phelps' work to comprise volumes on general topics followed by volumes of topography for every county, was diverted from its purpose by the upheaval of the First World War and was later saved by a combination of personal commitment, academic foresight and, in some counties including

Somerset, generous local-authority patronage. The *VCH* continues to combine the topographical awareness of Gerard and Strachey and the erudition of Dugdale with the assiduity of Harbin and Palmer in searching both local and national archives—together the necessary attributes of the county historian.

5
Early Topographers, Antiquarians and Travellers in Dorset

JOSEPH BETTEY

The first topographer to attempt a comprehensive study of the whole county of Dorset was Thomas Gerard (1592–1634), a country gentleman who lived at Trent near Sherborne. Earlier writers included John Leland, who visited parts of Dorset during the 1530s and left brief but informative descriptions of some of the towns, churches and former abbeys, notably Sherborne, Lyme Regis, Bridport, Beaminster, Melcombe Regis, Poole and Wimborne Minster. Leland also provided a useful account of the landscape, farming and people of the island of Portland, including the medieval castle and the newly built royal fortress, although curiously he says nothing of the quarries which were already an important feature of the island's economy.[1] Dorset is also described in a chapter of William Camden's *Britannia* of 1586, while the earliest printed map of the county was produced by Christopher Saxton in 1575, followed by William Kipp's map of 1607, and by the useful map published by John Speed in 1610, which provides much fuller information, including many place-names and the hundred boundaries, as well as the first plan of Dorchester.

For the first detailed description of the topography of the whole county, however, we are indebted to Thomas Gerard who produced his *Survey of Dorsetshire* during the 1620s, and who also wrote a description of the southern part of Somerset, *The Particular Description of the County of Somerset*, 1633.[2] The manuscript of Thomas Gerard's work on Dorset passed into the hands of his relatives, the Coker family of Mappowder, and when it was eventually published by the London bookseller, John Wilcox, in 1732 under the title *The General Description of the County of Dorset*, it was said to be the work of John Coker (d. 1635) who had been the rector of Tincleton near Dorchester. The true authorship of the work has, however, been made clear in a modern edition, and it is evident that the surveys of Somerset and Dorset are by the same person, while in his description of Trent, which was formerly in Somerset, Gerard wrote of 'the place now giving me my habitation'.[3]

The Gerard family had been settled at Trent since the early sixteenth century, and Thomas Gerard was the fourth of his line to live in the manor house. Like many of his contemporaries, he was extremely interested in the history and lineage of the local gentry families and proud of his own connections, decorating the transept arch of his family pew in the parish church at Trent with forty armorial shields showing the alliances of the Gerard family. Curiously, the inner face of the arch bears the text in mirror-writing, 'All Flesh is as Grass and the Glory of it as the Flower of the Field'. In 1618 Thomas Gerard had married Anne Coker, the 14-year-old daughter of Robert Coker of Mappowder in the Blackmore Vale, and this brought him into contact with many of the leading families of Somerset and Dorset, and he was thus able to gain access to collections of family papers and genealogies for his antiquarian research. Becoming a trustee of his father-in-law's estates, he had dealings with substantial Dorset landowning families such as the Strangways, Trenchards and Strodes, and he was also familiar with the work of antiquarian writers with similar interests in other counties.[4]

For his topographical survey of Dorset, Gerard used the river valleys as a way of describing the whole county, paying particular attention to the estates and houses of the gentry families. He included good descriptions of the small towns and markets of Dorset, and was also interested in the derivation of the place-names of the county. Like many later writers, he was struck by the contrast between the clay vales of the north and west of the county and the chalk downlands. Of the former he wrote that it was, 'abounding also with verie good Pastures and Feedings for Cattell; watered with fine Streames, which take their Courses through rich Meadows'.[5] Whereas he described the well-drained downland which, 'consisteth altogether of Hills . . . all overspread with innumerable Flockes of Sheepe, for which it yeelds very good and sound Feedinge, and from which the Countrie hath reaped an unknowen Gaine'. He also recognised that the economy of the county depended almost entirely upon farming and wrote that 'the more general Commodities of Dorsetshire are Cattell, Corne and Sheepe'. It is obvious from his descriptions that Gerard travelled extensively throughout the county in compiling his *Survey*. He was impressed by the strong fortifications of the castles at Corfe and Sherborne, both of which were later to be tested in the Civil War, and he was attracted by the splendour of the earl of Suffolk's new castle at Lulworth. He commented on the recent improvements which had been made to the meadows in the Frome valley around Dorchester, the beginnings of an innovation which was soon to develop into the great extension of water meadows throughout the chalkland valleys of the county, and he described the Frome passing 'amongst most pleasant Meadowes, manie of which of late Yeares have beene by Industrie soe made of barren Bogges'.

Gerard's tours of the county must have taken place during the early 1620s, for at Winterborne Anderson he wrote 'where of late Mr Tregonwell hath built a faire house', and this elegant building was completed by Sir John Tregonwell in 1622. Likewise, he described Up Cerne 'where Sir Robert Meller hath built a house', and Sir Robert Meller died in 1624.

The effects of the suppression of the monasteries were still very apparent in Dorset, and at Cerne Abbas Gerard noted that:

> The abbey came to a final period under Henry the Eighth who lay soe heavie upon Religious Houses that hee crushed all their substance into his owne Coffers . . . The Abbey Church is nowe whollie ruinated and onlie a small parte of the House standing, with a faire Tower or Gatehouse . . . The Towne which by the suppression of it lost its livelihood, and nowe remains (as most Abbie Townes) but poore, having noe certaine Trade for their Releefe; onlie still they enjoye a Weeklie Markett and some Faires.

Milton Abbas had also suffered badly from the suppression of its abbey and its acquisition by the Tudor lawyer and royal servant, Sir John

Tregonwell. Gerard noted the decline of the town and the sorry state of the abbey church, 'and that which remaines now serveth the Towne, which surelie is not the richest, having lost the chiefe Help, the Abbot'. Wimborne with its long history and great minster church he described as 'much more commendable for what it hath beene than for what it is'. Some towns were prosperous, however, notably the twin ports of Weymouth and Melcombe Regis where Gerard was impressed by the number of ships and the trade with Newfoundland, as well as the busy traffic with France. At Sherborne, Shaftesbury and Blandford Forum he commented upon the well-frequented markets, while of Dorchester he wrote that the town 'hath encreased and flourished exceedinglie, soe that nowe it maye justlie challenge the Superioritie of all this Shire, as well for quick Marketts and neate Buildings, as for the number of the Inhabitants; manie of which are Men of great Wealth'. As a perceptive traveller, Gerard did not fail to observe the evidence for the decline or total desertion of many of the villages along the chalkland streams. He also reasoned that such desertions were connected with the great expansion of sheep farming and was critical of those he held responsible. Thus at the former site of Winterborne Farringdon, south of Dorchester, where only the church of St German remained, he wrote: 'a lone church for there is hardlie anie House left in the Parish, such of late hath beene the Covetousnesse of some private Men, that to encrease their Demesnes have depopulated whole parishes'. Finally, further evidence that Gerard personally visited the places he described occurs in his account of Sir William Clavell's attempts to exploit the resources of his estate at Smedmore. Clavell had spent large sums in prospecting for alum and copperas, and had also established a salt works, using the oil-bearing shale from his estate as fuel to boil the sea-water. Later, Clavell had set up a glass works, again using the shale as fuel. Gerard described the shale as 'a kind of blueish Stone that serveth to burne for maintaining Fire in the Glasse House'. He also noticed a major disadvantage of this fuel, since 'in burning [it] yeelds such an offensive Savour and extraordinarie Blacknesse, that the People labouring about those Fires are more like Furies than Men'. Writers such as Gerard did not set out to produce histories in the modern sense, but their intimate local knowledge of people, families, places, buildings and customs means that their work is of great value for local historians and often provides unique insights into seventeenth-century conditions, attitudes, individuals and topography.

Two travellers of a very different sort visited parts of Dorset in 1635, and both left brief accounts of what they saw. The first was a Cornishman, Peter Mundy, who spent most of his life in travel, visiting many parts of Europe and undertaking a voyage to India and China.[6] Mundy came to Weymouth by sea from Falmouth. He visited Portland and described Chesil Beach and the Fleet with its many swans and wildfowl; he also makes an interesting reference to the attempt which was being made to drain part of the Fleet and turn it into pasture land. This totally impracticable scheme was the brainchild of a group of speculators led by the Horsey family of Clifton Maybank. Vast sums were spent on this abortive project, which led to the ruin of the Horseys and to the demolition of much of their house, and Mundy described the expensive sluices, trenches and drains which had been dug to remove the water and keep out the tide. 'But as yet all is to little purpose, the maine sea soakeing through the beach all alonge. It is sayd they will proceed afresh.' On Portland Mundy described the two castles, the rough water of the Portland Race, and the large quarries which had been opened to supply stone for the elaborate scheme devised by Inigo Jones for the repair of old St Paul's cathedral in London. He was also impressed by the lack of trees and consequently of fuel on the island, and by the fact that dried cow-dung was used for fuel just as he had seen it used in India. This practice continued on Portland and elsewhere in Dorset until the nineteenth century and was noted by several observers. Peter Mundy also visited Dorchester, seeing Maumbury amphitheatre and climbing up to Maiden Castle where, like so many others, he was greatly impressed by the huge earthworks 'of great use in Auntient tymes questionlesse'.

The other visitor in 1635 was Lieutenant Hammond who, with two companions, made a tour of the western counties from their home in Norwich.[7] The three travellers visited Badbury Rings, Wimborne Minster 'an ancient old Place', Poole and Corfe Castle where the military observers examined the strong fortifications. From Corfe they went to Wareham, where they noted the ancient burghal defences, and then to Lulworth Castle where they were lavishly entertained by the earl of Suffolk and greatly admired the recently completed castle 'stately and loftie, and newly built of Freestone . . . with 4 great and lofty high round towers at the corners'. They also visited Dorchester 'sweetly scytuated in a pleasant Valley', Maiden Castle, Weymouth with its crowded shipping, and the Island of Portland where they saw the castles and 'a great Companie of Prisoners, digging, delving, haling, breaking and framing stones to repayre that old, plaine, decaying and goodly Pile of Building, the Mother Cathedral of Our Kingdome'. Finally they visited Bridport and thence on to Lyme Regis and were impressed by 'the rare unparalleled Harbour called the Cobb' where they saw between thirty and forty ships protected by the stone wall of the harbour 'safely impounded and lockt up from wind and weather'.

Thomas Fuller, the author of *The Worthies of England* (first published in 1662), was vicar of Broadwindsor from 1634 to 1660, but his writings add little to our knowledge of Dorset, apart from his comments on the widespread cultivation of hemp in west Dorset where it provided the essential raw material for the rope, twine and net industry of Bridport and the surrounding area. Fuller wrote that 'England hath no better [hemp] than what groweth here betwixt Beaminster and Bridport, our land affording so much strong and deep ground for the same . . .'. As a clergyman it is curious that Fuller should have selected among his Dorset 'worthies' Arthur Gregory of Lyme Regis whose skill in forgery and in opening letters had played an important part in the downfall of Mary, Queen of Scots. This hardly admirable craft evidently impressed Fuller who described Arthur Gregory as having: 'the art of forcing the seal of a letter; yet so invisibly that it still appeared a virgin to the exactest beholder'.[8]

Later in the seventeenth century Dorset was visited by the enthusiastic observer and fearless horsewoman, Celia Fiennes. On Brownsea Island in Poole Harbour she described in some detail the production of copperas (sulphate of iron) which was used to produce a green colour for dyeing and tanning:

> They place iron spikes in the pannes full of branches, and so as the liquor boyles to a candy it hangs on those branches: I saw some taken up, it look't like a vast bunch of grapes, the coullour of the Copperace not being much differing it looks cleare like sugar-candy.

At Swannage she observed that the poor burnt the oil-bearing shale for both heat and light, but added 'it has a strong offensive smell'. She also visited Dorchester: 'the town lookes compact and the streets are very neatly pitched and of a good breadth, the Market-place is spacious, the Church very handsome and full of galleryes'. Finally she described Lyme Regis and the harbour within the protective wall of the Cobb.[9]

In contrast to Celia Fiennes, the journalist Daniel Defoe provides a more sober and measured though less lively account, especially of the economic life and farming of the county. Defoe's *Tour*, published during the 1720s, is written in the form of an itinerary which went first to Wimborne Minster, where he reported that 'the inhabitants, who are many and poor, are chiefly maintain'd by the manufacture of knitting stockings, which employs a great part indeed of the county of Dorset'. At Poole he observed the trade in oysters and on the Isle of Purbeck he noted 'the vast quarreys of stone'. Along the coast he commented upon the decoys for wild-fowl, especially the decoy at Abbotsbury 'the famous swannery, or nursery of swans, the like of which I believe is not in Europe'. Weymouth he described as 'a sweet, clean, agreeable town . . . and has a great many good substantial merchants in it, who drive a considerable trade, and have a good number of ships belonging to the town'. At both Weymouth and Lyme Regis, Defoe noted particularly the trade with France, Portugal, Spain, Newfoundland and Virginia. On the Island of Portland he saw the many stone quarries from which 'our best and whitest free stone comes', and elsewhere in the county he commented upon the manufacture of cloth, rope, sailcloth, silk and lace. He was particularly struck by the lace he saw at Blandford Forum, 'chiefly famous for making the finest bonelace in England, and where they shewed me some so exquisitely fine, as I think I never saw better'.

Above all, Defoe, like other travellers in the county, was amazed at the number of sheep which were kept on the downs, and at their crucial role in the farming economy for folding on arable land, as well as in the value of their wool, mutton and early lambs. As a dissenter, Defoe had few comments upon the churches, for example at Sherborne, noting only that 'the church is still a reverend pile and shews the face of great antiquity'. But he could not help being impressed by the friendly spirit, genteel society and abundance of ancient gentry families he found in Dorchester, and his eulogy on the attractions of the town is worth quoting at length.

> Dorchester is indeed a pleasant and agreeable town to live in, and where I thought the people seem'd less divided into factions and parties than in other places; for though here are divisions and the people are not all of one mind, either as to religion, or politicks, yet they did not seem to separate with so much animosity as in other places: Here I saw the Church of England clergyman, and the Dissenting minister, or preacher drinking tea together, and conversing with civility and good neighbourhood, like catholick Christians and men of catholick and extensive charity. The town is populous, tho' not large, the streets broad, but the buildings old and low; however, there is good company and a good deal of it; and a man that coveted a retreat in this world might as agreeably spend his time, and as well in Dorchester, as in any town I know in England.[10]

In 1730 the first narrative history of Dorset was published. This was *The Compleat History of Dorsetshire* by Thomas Cox, part of the author's six-volume *Magna Britannia* (1720–31). Although less than sixty closely-printed pages, Cox's work is densely packed with information. He provides a general description of the county, its major towns and administrative divisions, the markets, ecclesiastical benefices, gentry families and men of note born within the county. There is also brief historical information on the former monastic houses, boroughs, charity schools, landowners, descent of properties and major buildings.

During the eighteenth century increasing numbers of tourists visited Dorset and wrote accounts of their impressions of the county. Among them was Dr Richard Pococke, the Irish bishop who made extensive tours in England during the 1750s. In September 1750 he came from Salisbury across Cranborne Chase to Wimborne Minster, observing the numerous barrows and prehistoric earthworks along the route. He made detailed notes on the minster church at Wimborne and speculated on its long history, travelling on past Kingston Lacy to Badbury Rings, where he noted the massive defences and the Roman road to

Old Sarum. He also visited Poole and reported that, 'They have some Newfoundland trade, and a considerable business in building ships, and bringing the materials; they are also employed in fishing, having besides the common sea fish, plenty of soles and John Dory and very large oysters'. On the nearby heath he observed the stone quarries and the extraction of clay which was used for the manufacture of tobacco pipes. At Corfe Castle he saw the huge stone walls and fortifications which had been destroyed after the Civil War, and commented that 'when it was blown up by Oliver, the walls fell down in vast pieces'.

In October 1754 Bishop Pococke again visited Dorset and greatly admired the beauty of the countryside, especially the fertile lands of the Blackmore Vale and west Dorset, 'all an exceedingly rich and beautiful countryside'. On this occasion he also visited several of the largest mansions in the county, notably Wimborne St Giles, Bryanston, Eastbury, Milton Abbas and Wolfeton, making detailed notes on the situation, architecture and contents. He also walked around several towns. At Blandford Forum he wrote 'the town is well built, having been burnt down in Queen Elizabeth's time and again in 1731'; at Milton Abbas 'the town is a very small poor place'; in Dorchester he was greatly interested in the Roman remains, and at Sherborne he made detailed notes on the history of the town, its former bishopric and the Benedictine abbey. Particularly notable is Pococke's account of his visit to Cerne Abbas for he provides one of the earliest accounts of the Cerne Giant. He described the town of Cerne Abbas as: 'a large poor town, being near a mile in circumference. They make malt, and are more famous for beer than in any other place in this country; they also spin for the Devonshire clothiers.' Pococke made brief notes on the site of the abbey and the few standing remains which were visible. Finally he described the Giant:

> A low range of hills ends to the north of the abbey, on the west side of which is a figure cut in lines by taking out the turf showing the white chalk. It is called the Giant and Hele, is about 150 feet long, a naked figure in a genteel posture, with his left foot set out; it is a sort of Pantheon figure. In the right hand he holds a knotted club; the left hand is held out and open, there being a bend in the elbow, so that it seems to be Hercules, or Strength and Fidelity, but it is with such

> indecent circumstances as to make one conclude it was also a Priapus. It is to be supposed that this was an ancient figure of worship, and one would imagine that the people would not permit the monks to destroy it. The lord of the mannor gives some thing once in seven or eight years to have the lines clear'd and kept open.[11]

It would be of interest to know how Bishop Pococke was able to obtain so much detailed historical information during his brief stay in each place, but he gives us no clues as to his sources. No doubt some facts could be gleaned from local antiquarians and gentry or from the parish clergy, although these sources were not always reliable or accurate. The sort of informant upon whom some travellers had to rely can be illustrated from the experience of the archaeologist William Stukeley who visited Dorset in August 1723. His itinerary, with visits to many hill-forts, earthworks and barrows, may have been suggested by Lord Pembroke with whom Stukeley had stayed at Wilton. He visited Bokerley Dyke on Cranborne Chase, Coombs Ditch near Blandford Forum, Maiden Castle and Poundbury. He puzzled over the origin and purpose of the prehistoric field systems which were to be seen all over the chalk downlands, and wondered about the many lynchets to be seen along the hillsides. He spent some time in Dorchester, making detailed drawings and plans of the Roman amphitheatre known as Maumbury Rings. Later in 1723 he read an account 'Of the Roman Amphitheater at Dorchester' to the Freemasons' Lodge which met at the Fountain Tavern in the Strand in London. This was later published, together with five drawings, in his *Itinerarium Curiosum*. Returning to Wilton by way of Gussage All Saints, Stukeley stayed overnight at the Rose Inn, where the landlady told him a remarkable mixture of fact and fantasy about Roman remains in the district, the legendary seven churches of Knowlton, and stories of the deserted villages of which so many remains were visible on the chalklands, ending her saga with a ballad of Troy: 'Waste lie those walls that were so good / And corn now grows where Troy towers stood'.[12] Historical accuracy was not a high priority of such tales, and the situation encountered by Stukeley is reminiscent of Thomas Hardy's fictional account in *The Mayor of Casterbridge* of the conversation in 'The Three Mariners':

> Casterbridge is an old, hoary place o' wickedness by all account. 'Tis recorded in history that we rebelled against the King one or two hundred years ago, in the time of the Romans, and that lots of us was hanged on Gallows Hill, and quartered, and our different jints sent about the country like butcher's meat; and for my part I can well believe it.

After 1765 journeys in Dorset were made easier by the publication of a new and accurate map of the county by Isaac Taylor. This showed the major routes and many minor roads, as well as the towns, villages and most hamlets; it also contained fine drawings of Corfe Castle, Maiden Castle, Maumbury Rings, Sherborne Castle, Horton Tower, Lulworth Castle and a stone quarry on Portland.

Arthur Young provides a great deal of information about Dorset farming, livestock, farm buildings and agricultural improvement, although curiously this is included in his published account of *A Farmer's Tour through the East of England*, 1771. Young commented upon the importance of sheep in the farming economy of the county and gives a detailed account of the many water-meadows which had been developed to provide early grass for the flocks. He was scathing in his criticism of the Dorset practice whereby farmers avoided the daily chore of milking and the tasks of producing butter and cheese by renting out their cows to dairymen. Young noted that dairies were rented for £3 12s 6d per cow, and commented 'Was there ever such a ridiculous system known?'. He was also very critical of farmers for allowing the vast expanses of heath to remain unimproved. Clearly he failed to appreciate the difficulties of bringing these ill-drained, acid soils into cultivation and was over-optimistic about their potential. 'What a pity', he complained, 'that such extensive wastes should remain in so desolate a condition.'

Young was, however, full of admiration for some of the farms he saw and for the improvements which had been made by some estate-owners, notably by the Framptons at Moreton, the Sturts at Crichell, the Drax estate at Charborough, the Revd Dr Lloyd at Puddletown, Lord Milton at Milton Abbas and the Damer estate at Winterborne Came. Young also appreciated the beauty of the Dorset landscape, and his opinion is worth quoting, since he had travelled all over England

and had visited estates and farms throughout the country: 'A more varied or more beautiful country is no where to be seen in England than from Dorchester all the way to Bridport, and well worth a long journey to see'.[13]

During the later eighteenth century the number of tourists visiting Dorset increased rapidly as the aristocracy and gentry followed the example of George III, and began to take holidays and indulge in sea-bathing along the Dorset coast. It was to cater for such visitors that numerous topographical books on the county were published, such as J. Love, *The Picturesque Beauties of the County of Dorset* (1793); W. G. Maton, *Observations on the Western Counties* (1797); J. Britton and E. W. Brayley, *A Topographical and Historical Description of Dorset* (1810). The latter was part of the authors' *Beauties of England and Wales* (1803), and is especially valuable for its illustrations, as well as for the text.[14]

From 1774, however, tourists and writers on Dorset had the benefit of one of the fullest and best of all the great county histories, in the form of the massive *History and Antiquities of the County of Dorset* by John Hutchins, and it is appropriate to end this account with a brief notice of his remarkable achievement.[15] As soon as it was published, this work by a retiring Dorset clergyman became, and has remained, the essential starting point for any historical or antiquarian research on the county. John Hutchins (1698–1773) spent his whole career in Dorset. He came from a clerical family, and his father had been the incumbent at Bradford Peverell. After Dorchester Grammar School and Balliol College, John Hutchins was ordained, and in 1723 became a curate at Milton Abbas where he began the research for the project which was to occupy him for the rest of his life. He became rector of Swyre in 1729, and rector of St Mary's, Wareham in 1744. As a clergyman he presented a shy, reserved and scholarly figure to his parishioners, and although he was conscientious in his parochial duties, he was not an inspiring preacher and was ill-equipped to cope with the many dissenters in his parish at Wareham. Hutchins himself complained of the difficulties of visiting libraries and consulting sources, and of the interruptions to his historical work occasioned by 'constant attendance on a laborious cure', but nonetheless, encouraged and occasionally financed by some of the county gentry, he travelled throughout Dorset, and also to the Diocesan Registry at Salisbury and to record collections in Oxford, London and elsewhere, and his work is notable for its constant reference to, and long quotations from, original sources. Like all other eighteenth-century histories it was written about and for the gentry, and the first edition contained little on the poor or on economic history, but it is invaluable for its parochial histories with topographical and biographical detail, the descent of lands, ecclesiastical history, gentry houses, antiquities and for its illustrations. His work narrowly escaped total destruction in 1762 when his house and most of its contents was consumed by fire while he was away, but his notes and manuscripts were saved by his wife.

Sadly, Hutchins died in 1773, a year before the publication of his great work in two volumes. The editors of subsequent editions added considerable new material, especially on archaeology and social history, and the second edition of 1796–1815 was extended to four, large volumes, as was the third edition of 1861–74. The single-handed achievement of John Hutchins in producing such a detailed, scholarly and comprehensive county history provides a fitting culmination to this survey of early topographers and antiquarian writers on Dorset.

6

From Romanticism to Archaeology: Richard Colt Hoare, Samuel Lysons and Antiquity

MALCOLM TODD

After a very fruitful century, extending from the fieldwork of John Aubrey to the fieldwork of William Stukeley, antiquarian studies in Britain fell into a phase of relative lassitude.[1] The great days of topographical discovery seemed to be over. After 1754, it is difficult to name a single work of antiquarian scholarship which shows any real sign of innovation in thought or action. There is, of course, a distinguished exception and that is William Roy's masterly *Military Antiquities of the Romans in North Britain*, eventually issued in 1793 but relying mainly on observation and survey done in the aftermath of the '45 rebellion. The tradition of Camden was still alive, it is true, and his great *Britannia* was refurbished and greatly amplified by Richard Gough in his three-volume edition of 1789. But this book was effectively a failure, a dead-end, for it was now impossible to deal with all that antiquity comprehended in a single work; and the conceptual basis was changing fast. The fieldwork of the previous century had seen to that. Far more useful in the later eighteenth century were the studies of counties and regions, of which Nichols' *Leicestershire* and Borlase's *Cornwall* are perhaps the best and most enduring. The reasons for this relatively unproductive phase are not easy to define, but among them is without doubt to be numbered a shift in aesthetic taste which affected literature, art, architecture, manners, travel and much else, including how the remoter past was viewed. This change of outlook was immensely complex in form and its common designation as the Romantic Movement does not fully express what it was nor what it sprang from.

For the past sixty years, there has been a general acknowledgement of the proposition that the origins of modern archaeological thought, namely the study of Man through his works, from stone tools to cathedrals, lie in the period covered by the half-century from 1770 to 1820 and owe much to the Romantic Movement in literature and art. This idea has been given wide currency by the writings of Stuart Piggott in particular, virtually the only archaeologist to make a lasting contribution to this aspect of the history of ideas.[2] In so far as Romantic ideals were a reaction to rigid Classicism and the Enlightenment, the late eighteenth century certainly did see a major shift in how the human past was viewed. But there is little evidence for the beginnings of archaeological enquiry here. It is even far from clear that the intellectual climate was particularly favourable to the study of the material human past. I would prefer to adduce more down-to-earth reasons for the emergence of curiosity about the remoter past in these decades to either side of 1800. For there is no serious doubt that archaeology and related topography were transformed between 1770 and 1820. Those who brought this about were a relatively small company, among whom Sir Richard Colt Hore and the brothers Daniel and Samuel Lysons were perhaps the most effective partners and certainly the best known. Their work in turn relied upon a larger cadre of local antiquaries, most of whom languish in the shade and whose fieldwork and observation did not always find independent publication.

It is worth reminding ourselves of the prevailing state of knowledge about the remoter human past in the late eighteenth century. While the study of ancient Egypt, Classical Greece and Rome was developing within a more or less well defined framework of ancient history, the texts of which, both biblical and Classical, were known to every educated reader, prehistoric archaeology was still mired in a past which had neither chronology nor sequence. How could it? Ancient literary sources for Britain spoke only of ancient Britons as the people who were responsible for the sites familiar to readers of Stukeley's *Itinerarium Curiosum* or Grose's *Antiquities of England and Wales*. The world had begun in 4004 BC, according to Bishop Ussher; between that date and the first Classical writings all the megaliths, stone celts, circles, standing stones, loggan rocks, hill-forts, even Stonehenge and Avebury, had to jostle in time, assisted only by reference to Druids and their savage rites. The Romantic Movement was naturally much taken by Druids, as it was by hermits and other drop-outs. As Edith Sitwell put it: 'Nothing, it was felt, could give such delight to the eye as the spectacle of an aged person with a long grey beard and a goatish rough robe, doddering about among the discomforts and pleasures of Nature'. Not until the early years of the nineteenth century would this fantasy world be first questioned and then dissolved by increasingly scientific enquiry.

This preamble is necessary to set the scene for the start of the new age and to reveal more fully the achievement of the men who were to transform antiquarianism into archaeology in Britain between 1770 and 1820. The two leading figures were Sir Richard Colt Hoare and Samuel Lysons, men of different backgrounds and aptitudes, each making formative contributions which effectively cast the study of the visible past into an entirely new mould.

First, the work of Richard Colt Hoare. The son of Henry Hoare the highly successful banker, he inherited a massive fortune from his father, along with Stourhead in Wiltshire. He was a man of discerning taste as well as a scholar (the two are not always close companions) and his mind was distinguished by its independence as well as its range. Although possessing vast inherited wealth and not needing to lift a finger to obtain the necessities of life, he was industrious and devoted most of his life to travel, learning, fieldwork and writing.[3] He was the central figure in a large network of scholars, good, bad and indifferent: William Cunnington, John Skinner, Philip Crocker, John Carter, Thomas Leman. The most significant of these was the fieldworker Cunnington. Colt Hoare's great work was to examine a huge number of the barrows and other monuments on Salisbury Plain and to publish the results of these excavations. It was a research programme, so far as the time could have produced such a thing, and it was as scientific as the age could aspire to. The publication of the results was even more impressive than the work. In 1812 there appeared the first volume of *Ancient Wiltshire*, the most magnificent excavation report to have appeared down to that date. It opens boldly: 'We speak from facts, not theory. Such is the motto I adopt and to this text I shall most strictly adhere. I shall not seek amongst the fanciful regions of romance an origin for our Wiltshire Britons.' Romanticism was out, or at least on the way out, and the Druids in retreat, though they could still make a final twitch on the thread. *Ancient Wiltshire* had great and almost immediate *reclam*, with the reading public as well as with the learned world. The *Quarterly Review* published a long notice of the first two parts of the book, favourable on the whole, though commenting on the introductory theory as 'inconsistent with observed facts'.[4] It would be interesting to know who felt secure enough to make this observation. Oddly, the typography was singled out for especial mention as being 'perhaps unrivalled'. But there were more percipient comments, as on Colt Hoare's treatment of Stonehenge. 'We have the satisfaction of finding that he has conducted himself with all the sobriety, modesty and discretion which become a modern antiquary in treating a subject of such difficulty, in which so many of his forerunners have failed.' One has to add that it has scarcely been possible to pass such a judgement on any treatment of Stonehenge since then. The overall critique also seems reasonable:

> It is a striking proof either of his own influence, or of the neglected state of the country, that he has everywhere been allowed to prosecute his researches with as little interruption as if he had been digging on his own estate. No antiquary had ever the same means or opportunities before Sir Richard Hoare; and no-one ever availed himself more entirely of the advantages which he possesses. In his knowledge of barrows, he certainly stands unrivalled. He has reduced the subject to a system and has nearly invented a technical language in which to describe it.

Although sales of the book were disappointing to Colt Hoare and he lost heavily on its production, the fame of the publication was immediate and not merely in England. Far away in Weimar, a copy found its way into the library of Goethe, himself much engaged in prehistoric studies at this date.[5] The notice in the *Gentleman's Magazine* for 1811 was much more complimentary and has the added spice of an attack on the piece published in the *Quarterly Review*. The author of the latter is roundly attacked as grossly unfair to William Cunnington and pathetically ignorant of the realities of fieldwork:

> Now I strongly suspect that our learned Critick is not quite so well informed on this matter as he ought to be. I doubt if he has ever seen a Barrow opened—or explored a British village: perhaps also he has never traversed the Wiltshire Downs, with a view of inquiry into their Antiquities.[6]

One wonders who the author of this review was. He may not have been a member of Colt Hoare's immediate circle. Its pugnacious style and partisanship for observation (and action) in the field are reminiscent of Samuel Lysons but this cannot be corroborated. Despite the difficulties which attended the making of *Ancient Wiltshire*, which involved dissension between the antiquaries and fieldworkers who gathered together at Stourhead, there is no doubt that its publication marked a decisive break with the fanciful past. While it was being prepared, the influence of other practical intelligences was at work on the study of the British past, chief among them those of the brothers Daniel and Samuel Lysons, to whom we now turn.

The Lysons brothers came from a lesser social world than that of the master of Stourhead. Sons of a Gloucestershire clergyman, they had to make their own living. Samuel, born a year later than his brother in 1763, attended Bath Grammar School and then embarked on a successful legal career. By 1796 he was enjoying connections at Court and in 1803 he was appointed Keeper of Records at the Tower of London. Precisely how his interests in the visible past had been aroused is not certain, but by the early 1790s he had become an active and forceful Fellow of the Society of Antiquaries; he was elected a vice-president and director in 1798.[7] It may have been his interest in painting, which he pursued with skill and flair, which first led him to ancient sites and objects. However that may be, one aspect of the past which attracted him early on was the discovery of Roman villas and their mosaics, first in his native Gloucestershire, later in Sussex and elsewhere. Remains of this kind had attracted a great deal of attention earlier in the eighteenth century, not least through the splendid engravings of George Vertue; but rarely had these discoveries been fully recorded and discussed. Lysons placed the subject of Romano-British villas and their adornments on an entirely new footing, partly by his own excavations but chiefly by his sumptuous publications, of which the most magnificent was the *Reliquiae Britannico-Romanae* which came out in two superb folio volumes between 1801 and 1817. Only fifty copies were printed and it is easy to see why. These were artistic products of outstanding quality, each set costing the subscriber £48 6s. The *Reliquiae* were a great achievement and a costly one. Lysons paid out £6,000 on their production. In the long run, however, more important were to be Lysons' own excavations, at Withington, Great Witcombe and Woodchester in Gloucestershire, and at Bignor in Sussex, in particular. These were the first excavations to reveal villa sites on a large scale and Lysons revealed a remarkably mature understanding of their plans. At Woodchester and Bignor he had the good fortune to find well-preserved mosaics which he recorded with accuracy and taste. At Woodchester, he hit the jackpot when he found the immense Orpheus mosaic flooring the great hall, the largest mosaic known in Roman Britain and one of the largest in western Europe. More impressive was Lysons' recognition of the scale and importance of the Woodchester villa, which he reasonably saw as the palatial residence of a high official. His publication of the Woodchester discoveries, *An Account of Roman Antiquities discovered at Woodchester in the County of Gloucester*, which appeared in 1797 with a dedication to George III, is a sumptuous volume, with a text in both English and French. The three beautiful coloured drawings of the site and its environs are evocative recreations of the late eighteenth-century setting. More workmanlike, but still very handsome, are the location map, the plan of the villa and the detailed coloured drawings of structural detail, mosaics, painted plaster, sculptures and small objects. There are even brief notices of small finds such as coins, a rare inclusion at this date. At Bignor, his excavation from 1811 to 1815 could be more extensive and the resultant plan was the best and most complete of a Roman villa in Britain down to that date.[8] The mosaics here too were illustrated charmingly and fairly faithfully. But Lysons' range was already much wider than villas and their furnishings. In 1804 he had published *A Collection of Gloucestershire Antiquities*, in which he illustrated and discussed churches, substantial houses, monuments, sculpture, stained glass, brasses and seals. The plates, we are informed, were etched from Lysons' own drawings.

The Gloucestershire volume was a harbinger of Lysons' most ambitious work, the *Magna Britannia*, undertaken jointly with his brother Daniel. This was to be a masterwork of topography, recording England county by county in such detail as had never before been attempted. Printed and manuscript records were to be used, as well as observations in the field by Lysons and others. Following the established practice of eighteenth-century antiquaries, questionnaires were issued to relevant correspondents and the replies collated.

The *Magna Britannia* was intended to be the most extensive topographical description of Britain attempted to that date. The first volume (of 1806) contained Bedfordshire, Berkshire and Buckinghamshire, the second Cambridgeshire and Cheshire, the third Cornwall, then Cumberland and Derbyshire. The sixth volume, on Devonshire, proved to be last. Samuel Lysons died in 1819 and his brother found the task of producing the Devon volume very heavy. Material for other projected volumes had been amassed in bulk and is still stored in the British Library. In the Devon volume the well-tried format was adhered to, brief introductory sections on the ancient inhabitants, ecclesiastical history, towns and markets, nobility and gentry, natural history and ancient sites being followed by the substantive body of the work, the parochial history, arranged in alphabetical sequence of parishes. Many of the individual parish entries are excellent, some are mediocre, a few are poor—perhaps a reflection of the removal of Samuel's guiding hand. But the best work is by far the best topographical writing on the county done down to that time. It was in fact the first substantial collection brought together since those of William Pole, Tristram Risdon and Thomas Westcote in the early seventeenth century. Parts of the county, and particularly Exeter, had been the subject of attention by others, but none of the resultant publications were of outstanding quality and several never got as far as publication at all. The Lysons' volume was far more ambitious than any of its forerunners, not least as it displayed a historical perspective lacking in its immediate predecessors and was firmly based on both documentary sources and topographical observation. It is not without faults and weaknesses. There is too much on families and their interconnections, and too frequent reference to individuals who no doubt appeared worthy at the time, but whose significance is now difficult to divine. But there is much of enduring value and certain passages have a surprisingly modern air and are recognisably akin to entries in early volumes of the *Victoria County History*. A few entries are even reminiscent of early Pevsner volumes and in some respects Lysons' work may be seen as a linear ancestor of the Pevsner series, albeit a remote one.

Lysons made little of Dartmoor and seems to have seen little of the rich antiquities of the region. But this passage on Grimspound clearly rests on autopsy:

> It consists of a circular enclosure of about three acres, surrounded by a low vallum of loose stones, some of which are very large, being the remains of a wall. There are two entrances opposite to each other, directly north and south; at these points the wall, which appears to have been about 12 feet high, were the thickest. Within the enclosure are numerous small circles of stone, in general about 12 feet in diameter; the greater part are near the south side of the enclosure. Various conjectures have been formed respecting this remarkable remnant of antiquity: some have supposed it a place of religious worship, others the remains of a British town, and connected with the ancient tin-works, the vestiges of which are visible near the spot.[9]

Given that Grimspound was not to be placed in any archaeological context for another eighty years, this is a reasonable account. The absence of any reference to Druids and their hocus-pocus is notable.

Contributions by other observers were included in the Devonshire volume, William Bennet the bishop of Cloyne providing much on the early antiquities of the county. His remarks on Hembury were to prove prophetic:

> If Hembury be not regarded as Moridunum, I am inclined to allow it to have been a British camp occupied by the Romans; it is an irregular figure, but tending to circular. Oval stones, used by the Britons for slings, have been found in it, yet its lofty situation, commanding the Vale of Otter, the ancient roads running up to it, the marks of two raised hills within the area, and some possible marks of occupancy, the

> Roman lar, and it is said coins dug up near it, with its very convenient distance from both Exeter and Seaton, are strong proofs of its having been possessed by the latter people.[10]

There are fortunate guesses in this, but reasonable deduction too. The main point, that Hembury was a prehistoric fortification later used by the Roman army, was demonstrated as a fact by excavation in the early 1980s.[11]

On other ancient works there are brief notices which are still worth remark. On Woodbury Castle, for example, Lysons records that military camps were sited here during the alarm of French invasion in 1798 and again in 1803, and that the ancient fort was occupied by an artillery park. It may be worth raising the possibility that some of the large, regular intrusions observed in excavation here in 1971 belong to this episode.[12]

One of the most interesting and innovative features of the Lysons' volumes as a whole is the emphasis upon medieval architecture, sculpture and single objects. This clearly reflects Samuel's experience at the Society of Antiquaries, to which notices of major discoveries and curiosities were sent by correspondents, read to the assembled *savants* and often subsequently published in *Archaeologia*. What is new is the attention to detail in Lysons' work, drawn out in the excellent engravings, and the sound judgement usually displayed (Figure 6.1). The romanticism of many eighteenth-century illustrated accounts has been replaced by a sharp but sympathetic realism. This approach to medieval antiquities could, and probably should, have led to the emergence of medieval archaeology. Why it did not is far from clear. Even the great restorations of churches in the mid-nineteenth century did not result in full analytical study of the material remains of medieval Britain. In their analysis and interpretation of medieval structures, and not least in their skilful illustrations, the Lysons' volumes were many decades ahead of their time and should have excited emulation. One view of the *Magna Britannia* is as a false start for medieval archaeology in Britain. Rickman's pioneering work on architectural styles likewise failed to give the impetus in this direction that might have been expected.[13]

After the *Magna Britannia*, even in its incomplete state, nothing was the same again in topographical studies. The mantle of the Lysons brothers passed to no single pair of shoulders. Henceforward, the great

Figure 6.1 Compton Castle, Devon, drawn by Samuel Lysons for the *Magna Britannia*, *Volume 6: Devonshire*, 1822.

tradition was continued by the Ordnance Survey, and later the *Victoria County History* and the Royal Commissions on Historical Monuments. That tradition is thus still alive, despite the indifference and occasional hostility of officialdom. As the end of the millennium approaches, it is far from clear that it will survive as a central element in the study of the British past.

We need to ask why the period we have been examining was so formative for topography and fieldwork. It behoves us to look not only, or principally, at philosophical or theoretical developments; much more banausic explanations often underlie major shifts in ideas and trends in the development of a subject. The parliamentary enclosures of the later eighteenth century not only struck a heavy blow at what still survived of the peasant culture of southern England. They also brought drastic alteration to the landscape in large tracts of the country. We know from our own experience of the past decade just how rapid such change can be. In concert with increasing industrialisation and the concentration of land in estates, the enclosures of the period from 1770 to 1820 altered the face of the land, destroying many earlier landscapes and sites and revealing much that had been hidden. Probably there had never before been such an upheaval which had wrought so much disturbance of the land. Many economic historians now play down the impact on the human population which earlier generations deplored; but the effects on the landscape were formative. They can scarcely have been lost on a considerable landowner like Colt Hoare and an acute observer like Samuel Lysons. Elsewhere in Britain at this time more account was being taken of antiquities in the landscape and, I would argue, for the same reason. Works of earlier settlers were being revealed as never before, and often being annihilated at the same moment. The stimulus to record and preserve, if possible, was powerfully present and in several areas of Britain the challenge was met. There is a large subject here waiting to be thoroughly explored.

What is beyond doubt is the fact that Colt Hoare and his circle and the Lysons brothers raised topographical study on to a new plane and this was not matched by anything in continental Europe. Henceforward, antiquities and their place in the landscape were seen in a completely different way. Romanticism had given way to realism and there was to be no turning back.

Appendix

The Illustrations in the Magna Britannia: the contribution of Charles Alfred Stothard (1786–1821)

Jane Marchand*

The illustrations for the *Magna Britannia* add greatly not only to its value as a work of scholarship on topography and antiquity, but also to its aesthetic appeal. The considerable artistic talents of Samuel Lysons had already been demonstrated in his published work before the *Magna Britannia* project was conceived, and inevitably his skills were again employed in providing illustrations for the latter. However, a variety of other notable artists were also engaged to produce drawings for the *Britannia* volumes (Figure 6.2). One of the most significant was the antiquarian draughtsman, Charles Alfred Stothard, born in London, the second son of the artist and illustrator, Thomas Stothard, RA, friend and collaborator of William Blake.

Charles Stothard had first made his name as a medieval antiquary with the publication of his *Monumental Effigies of Great Britain*, a work in which he depicted the changes in English costume from the twelfth century to the reign of Henry VIII, and thereby demonstrated his ability to produce exquisite drawings which were strictly accurate in historical detail. *Monumental Effigies* comprised twelve parts, the first ten by Stothard himself; the last two were published after his death.

In 1815 Stothard was employed by the Lysons brothers to produce drawings for the *Magna Britannia*, for which he travelled throughout the north of England. During his absence on this early commission, clearly impressed by the quality of his work, Samuel Lysons recommended Stothard for the position of historical draughtsman to the Society of Antiquaries. By 1816 he was therefore busy working on drawing projects for the Society and it was to be five years before he was asked by Daniel Lysons to make a further artistic contribution to the *Magna Britannia*. Daniel was by then engaged upon the task of completing the volume devoted to Devonshire and his invitation to

---

*Jane Marchand is currently researching 'archaeological thought and practice on Dartmoor in the nineteenth century' for the degree of MPhil at the University of Exeter.

Figure 6.2 View of Sidmouth drawn by J. Farington, RA and etched by Letitia Byrne for Daniel and Samuel Lysons' *Magna Britannia, Volume 6: Devonshire*, 1822.

Figure 6.3 The painted glass east window of Bere Ferrers Church drawn and engraved by R. Stothard for Daniel and Samuel Lysons' *Magna Britannia*, *Volume 6: Devonshire*, 1822. This is the illustration on which Charles Stothard was working when his fatal accident occurred. It was completed by his brother, Robert.

Stothard to assist him in the task fulfilled a promise he had apparently made to his brother, Samuel, before the latter's death in 1819. Thus, in 1821, Stothard visited Daniel at Rodmarton in Gloucestershire to arrange an appropriate itinerary which was to enable him, for example, to make drawings of the effigies in Exeter cathedral and the stained glass east window at Bere Ferrers. He undertook his journey through Devon on foot, his usual mode of travel, as the leisurely pace enabled him to stop and observe in detail the churches and monuments he encountered on the way.

Stothard is known to have arrived at Bere Ferers on 27 May 1821 and the next day began his drawing of the east window, which depicts the thirteenth-century founders of the church, Sir William and Lady Ferrers (Figure 6.3). He had, it seems, almost completed the task when a rung of his ladder broke and he fell, sustaining a fatal injury when he hit his head on the tomb of Sir William Ferrers at the north side of the altar. A brass plate in the chancel floor marks the place where this accident occurred and a memorial stone is affixed to the outer wall beneath the east window. Though now virtually illegible, the stone bears a lengthy eulogistic inscription to an artist whose life was so suddenly and prematurely cut short, but whose work had so significantly enhanced and enriched the pages of the Lysons' magnum opus. It is fortunate, however, that Charles Stothard's younger brother, Robert, also a talented artist, was able to complete the commission for the drawings in the Devon volume of the *Magna Britannia*.

7

The Scientific Gaze: Agricultural Improvers and the Topography of South-West England

SARAH WILMOT

The Rise of Agricultural Topography and its Methods

The collection of information for regional agricultural topographies formed part of the growth of interest in the study of nature, society, and the economy that arose at the end of the eighteenth century and during the nineteenth century.[1] Fact-gathering on the existing condition of customary agricultural practices was conceived as an important first step in promoting a more 'rational' approach to the exploitation of the capabilities of the land. This was not a new idea in the eighteenth century, but had its origins in the mid-seventeenth century Georgical Committee of the Royal Society, which promoted the study of regional agricultures so that 'it might be knowne what is knowne and done already, both to enrich every place with aides, that are found in any place, and withall to consider what further improvements may be made in all the practice of husbandry'.[2] Later surveys instigated by the Board of Agriculture (founded 1793), and the Royal Agricultural Society of England (founded 1838), shared these objectives and broadly reflected Enlightenment philosophy with its tendency to emphasise 'reason' and 'universal' principles in preference to tradition and regional diversity in agriculture. The hallmark of the scientific gaze was an emphasis on analysis, experiment and observation.[3]

William Marshall and Arthur Young were the two most prominent exponents of the new agricultural topography in the eighteenth century, and can be said to have been influential in determining the overall pattern of the subsequent regional surveys published by the Board of Agriculture. Unlike the majority of later agricultural writers, they were very explicit about the methods they employed in compiling regional descriptions. Additionally, historians have from time to time reappraised their methods in detail.[4] Arthur Young described his task as 'describing the husbandry of the Kingdom, by registering minutes on the spot', an undertaking 'never having been executed either in this or any other country in Europe'. On his first tour, begun in 1768, Young developed a questionnaire which formed the basis for collecting information in his subsequent regional surveys. In addition to keeping a daily register of observations, Young interviewed and cross-questioned local agriculturalists. As his reputation grew, Young was able to gain interviews with a wider circle of gentry, bailiffs and tenant farmers than was the case in his early travels.[5]

William Marshall used what he called the 'analytic method' of gaining agricultural knowledge on his tours, which involved the recording of immediate observations and the production of a 'systematic retrospect' at the end of each journey. This form of 'intentional observation', backed by scientific knowledge and experimental discovery, was the cornerstone of his approach. Marshall objected to those who wrote agricultural accounts based on the descriptions of local farmers without corroboration from direct investigation. Although Marshall frequently remarked that direct experience of farming in the region was the best qualification for making agricultural observations, this did not mean that local knowledge was necessarily of value if it was also uncritical of customary practices, and was not systematically observed.[6]

As an example of Marshall's 'on the spot' methods, the following extract is taken from his *Rural Economy of the West of England*, describing the soils of the South Hams district.

> IVYBRIDGE TO KINGSBRIDGE. The soil uniformly fertile. The tops of some of the hills are rich grazing ground! Other hills are leaner and less productive. But I observed not a field worth less than ten or

> fifteen shillings an acre . . . the hill sides are excellent corn land; . . . the bottoms rich meadows. Some little red soil is seen, in this ride. KINGSBRIDGE TO TOTNESS. The nature and appearance of the country are much like those observed between Ivybridge and Kingsbridge; excepting a high swell or swells, the soil of which is inferior to any, in the foregoing ride: the produce furzey, inclinable to heath: one of the Chudleigh Hills thrown in here. Much red soil appears in this ride. The water of the road, in some places, red almost as blood.[7]

John Barrell has commented of this form of description that it embodies no local knowledge at all, but builds up generalisations based only on what can be observed by the stranger riding through the district, and from general agricultural knowledge.[8] This is a common feature of the agricultural topographies written during the eighteenth and nineteenth centuries, although the balance between direct observation, reports of interviews with local agriculturalists, and direct farming experience varies considerably between the individual writers concerned. The balance also shifts within the work of any one individual, as regions are traversed with which the writer has differing degrees of familiarity.

Following these methods of obtaining information, agricultural reports represented simultaneously an enumeration of regional resources and manuals for the promotion of further economic development. The objectives set by the Board of Agriculture for their regional surveys are reproduced in the preface to John Billingsley's *General View* of Somerset. These were to examine:

> 1. The riches to be obtained from the surface of the national territory.
> 2. The mineral or subterraneous treasures of which the country is possessed.
> 3. The wealth to be derived from its streams, rivers, canals, inland navigations, coasts and fisheries.
> 4. The means of promoting the improvement of the people in regard to their health, industry, and morals, founded on a *statistical* survey.[9]

Eight reports on the South West were compiled for the Board of Agriculture: Dorset was surveyed by John Claridge in 1793 and W.

Stevenson in 1812; Devon was covered by Robert Fraser in 1794 and Charles Vancouver in 1808; Cornwall was covered by Robert Fraser in 1794 and G. B. Worgan in 1811, whilst John Billingsley was responsible for two reports on Somerset which appeared in 1794 and 1797.[10] In addition to these 'official' reports, William Marshall produced an independent survey of the *Rural Economy of the West of England* (Figures 7.1 and 7.2) in 1796.[11] Later surveys were much more narrowly agricultural in their scope and objectives, but they shared a similar pattern and ethos. The 1830s saw the appointment by government of assistant tithe commissioners whose job was to survey the agricultural potential of the region on a parish-by-parish basis.[12] Royal Agricultural Society reports appeared on Cornwall in 1845 and 1890, on Devon in 1849 and 1890, on Somerset in 1850, and on Dorset in 1854.[13] Common to all these surveys was a belief in the value of the systematic collection and co-ordination of regional information within a central administrative body.

Methods of collecting information were sometimes the subject of dispute between contemporary agriculturalists. William Marshall was a fierce critic both of Arthur Young's work and of the surveys produced by the Board of Agriculture. In Marshall's view, Young's lack of practical farming experience in most of the regions he toured led to an over-reliance on replies to questionnaires and a neglect of careful, systematic observation. Of the Board of Agriculture reports, Marshall's principal criticism was that they were compiled on a county basis, rather than on the basis of farming regions, which, like the Tamar valley, often straddled county boundaries.[14] He was also extremely critical of the abilities of many of the individuals chosen to undertake the reports, and in 1817 published a detailed assessment of each one in a five-volume work entitled *The Review and Abstract of the County Reports to the Board of Agriculture*.[15]

The Review and Abstract is an extremely useful document for attempting to assess the reliability of the agricultural surveys that were produced for the South-West region; however, it must be used with care. For example, Marshall's blanket condemnation of Young stands in need of correction in the light of R. C. Allen and C. O'Grada's recent findings. In their view, Young's data can be used with confidence, even if his data do not always support the conclusions that

THE

RURAL ECONOMY

OF THE

WEST OF ENGLAND:

INCLUDING

DEVONSHIRE;

AND PARTS OF

SOMERSETSHIRE,

DORSETSHIRE,

AND

CORNWALL.

TOGETHER WITH

MINUTES IN PRACTICE.

By Mr. MARSHALL.

IN TWO VOLUMES.

VOL. I.

LONDON:

Printed for G. NICOL, Bookſeller to His Majeſty, Pall Mall;
G. G. and J. ROBINSON, Paternoſter Row;
and J. DEBRETT, Piccadilly.

M,DCC,XCVI.

Figure 7.1 Title Page of William Marshall's *Rural Economy of the West of England*, 1796.

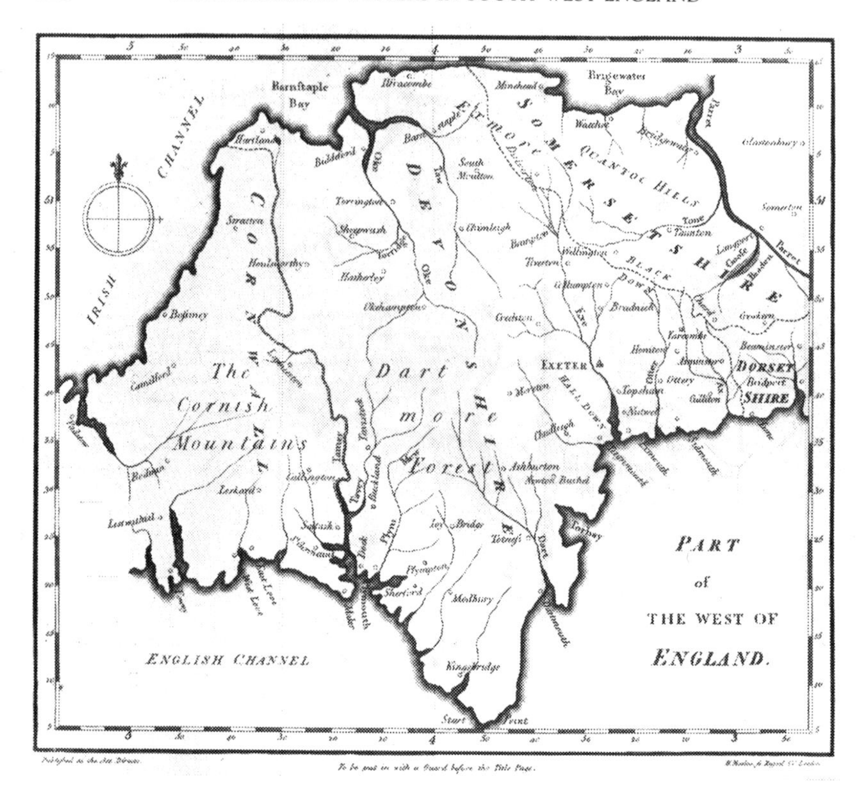

Figure 7.2 Map of Part of the West of England from William Marshall's *Rural Economy of the West of England*, 1796.

Young himself drew from them.[16] Marshall's opinion alone is thus not necessarily a safe guide to the value of every aspect of an agricultural reporter's work. However, Marshall does steer us towards some general principles for assessing the authenticity of eighteenth and early nineteenth-century agricultural surveys, and provides some insights into the credentials of individual agricultural observers.

The general principles of assessment include, firstly, the nature of the authorship of the essay; secondly, the local knowledge of the author; thirdly, the geographical coverage of the tours on which the essay is based; and fourthly, the quality of the author's contacts with local sources of information, principally the local farming community. In the first case, the question includes both the credentials of the individual concerned, and the issue of multiple authorship, for several of the Board of Agriculture reports were compilations from earlier reports.

The local knowledge of the author varies according to whether the individual is a resident of the county, with or without practical farming experience, or a regular or occasional visitor to the region. The geographical coverage of the essays is usually self-evident; often what purports to be a county survey is in fact a detailed survey of one or two districts only, with a general overview of the remainder. Finally, where much of the information is derived from interviews, it is important to be aware that these may over-concentrate on certain groups: the gentry and the improving farmers, for example, may be over-represented and it may be difficult to reconstruct a picture of average farming. In general, the essays should not be relied on in isolation for historical information but must be cross-checked with other contemporary sources.

Marshall's detailed assessments of the Board of Agriculture reports for South-West England are as follows. Of Robert Fraser's survey of Devon, Marshall notes that its coverage is patchy, being reasonably reliable for the South Hams and the Forest of Dartmoor, but so poor on north and east Devonshire, and the Exe Vale, that it 'ought to have been suppressed'. Of Captain Fraser himself, Marshall remarks that his 'education, early habits, and turn of mind' utterly unsuited him for the task of an agricultural survey. He is similarly dismissive of Fraser's essay on Cornwall, describing it as based on information collected in one winter season and 'hastily sent in' for publication.[17] G. B. Worgan's report on Cornwall is an example of multiple authorship. Marshall tells us that Worgan's original report was so unsatisfactory that the Board of Agriculture appointed three natives and inhabitants of Cornwall, of 'long experience in practical agriculture', to revise it. The three men, Robert Walker, Jeremiah Trist and Charles Vinicombe Penrose, were in Marshall's view probably amateurs of a superior class rather than 'professionals'. Overall he was complimentary about the final report that resulted.[18] Marshall's criticisms of Vancouver's survey of Devon mostly resolve into questions of literary style and organisation. He also draws attention to Vancouver's evident concentration on the agriculture of north Devon, and is critical of Vancouver's attempt to map the soils of the county.[19]

The first Board of Agriculture report for Dorset was produced by John Claridge, a partner, if not a pupil, of the royal land agent Nathaniel Kent. Although he was a knowledgeable estate agent, he was without agricultural experience and Marshall complained that his worst defect was 'want of digestion'. Marshall was even less complimentary regarding W. Stevenson's subsequent survey of Dorset. This report combined Stevenson's own observations made in 1811 with a manuscript report compiled by a Mr Batchelor in 1810. Consequently Marshall argued that Stevenson was the editor rather than the author of the report, 'and as such he only succeeded in making confusion worse confounded'.[20] The Somerset reports were completed by John Billingsley, who was a drainer and enclosure commissioner from Ashwick Grove near Shepton Mallet. Marshall conceded that he was a superior amateur with personal knowledge of drainage works in the 'western waterlands', but by implication not someone fully qualified to write a county-wide agricultural survey.[21]

Marshall's own study of the agriculture of the South-West region was the result of four visits to Devon during the summers and autumns of 1791–3 and 1795. His principal base was Sir Francis Drake's estate at Buckland Abbey, where he was advising on estate improvements. From there he made tours to Cornwall, Somerset and Dorset.[22] In the light of this it is apparent that Marshall's practical farming experience in the region was itself limited, and that the detailed coverage of the region in his *Rural Economy* was inevitably uneven. It is necessary to apply the same critical approach to interpreting Marshall's writings as is found to be necessary in appraising the work of Young and the Board of Agriculture reports.

These same general principles also apply to the assessment of the reports compiled for the Royal Agricultural Society during the mid to late nineteenth century. In the absence of a contemporary critic akin to Marshall for this period, the researcher is dependent on a much wider range of biographical sources to gain an insight into the backgrounds of the authors of these surveys.[23] These later surveys were produced in response to open competition for essay prizes offered by the Royal Agricultural Society, and judged by committee within that society, whereas the Board of Agriculture reporters were selected on the basis of patronage. In general one can assume a far greater degree of local knowledge for the Royal Agricultural Society reports, although as writers they were still often a very different breed from the 'practical farmers' of the region. Thomas Acland's report on Somerset, for example, was based on direct knowledge of the management of land on his family's extensive estates. This knowledge was supplemented by travels all over the country visiting landlords, farmers and agents. Three years earlier Acland had also spent a period at King's College, London studying chemistry with a view to improving his understanding of agricultural fertility.[24]

During the course of the production of these agricultural surveys, a new concept of the 'agricultural region' gradually emerged. Hugh Prince has remarked that writers increasingly moved away from the early eighteenth-century notion of regions as 'cultural entities steeped in local traditions'. In its place was the view of the agricultural region as a 'functional unit' whose 'farming practices were finely adjusted to physical and economic constraints'.[25] In search of systematic knowledge, agricultural writers incorporated insights drawn not only from wider agricultural experience, but also from political economy, geology and meteorology. As the nineteenth century progressed, this background information, increasingly available in statistical tables and maps, structured the arrangement of the county reports. Geological maps, in particular, shaped the descriptions of the agricultural topographers, although, as Prince has commented, geology was not everywhere a useful guide to agricultural diversity.

In an important sense, agricultural topography was never a simple description of place, but was a form of natural description which set the region within the context of a much wider body of knowledge, both practical and scientific. Agricultural writers also expressed a strong set of values concerning land and the uses to which it was put, against which regional agricultures were measured. If we are to understand the agricultural topographies of South-West England we must enquire more closely into the nature of these values. The following discussion therefore begins by examining what agriculturists deemed to be *ideal* topographies, speaking from the viewpoint of agricultural development. The reactions of agriculturalists to the topography of the

South West is then explored in the light of these improvement ideals. Agricultural topographers were also influenced by contemporary theories of beauty in landscape. Since these theories implied an increasing attraction to landscapes which resisted cultivation, the discussion concludes with an exploration of how, if at all, agricultural topographers were able to reconcile the conflict between ideas of utility and beauty when confronted with the topography of the South West.

Topographies of Improvement: The Ideals

To increase agricultural output and productivity was the avowed aim of every agricultural improver. In effect, a complete transformation of the rural landscape was envisaged, central to which was the disappearance of common land, the reclamation of unproductive 'wastes', the expansion of arable, the consolidation of farms, and the creation of new farm buildings and larger fields. Most prescriptions for agricultural improvement included these ingredients during the eighteenth and nineteenth century, although pressure to cultivate marginal lands evaporated with the onset of a depression in agricultural prices in the 1880s. Most of these changes in the landscape of farming also necessitated changes in the social structure of the farming community, a movement that can be loosely summarised as involving the emergence of a tenantry farming larger farms with increased capital.[26]

Agricultural improvers turned their prescriptions for agricultural progress into a *moral* discourse in their writings. Arthur Young, for example, declared in his *Farmers' letters to the people of England* (1771) that a 'waste acre of land is a public nuisance'.[27] John Beasely, ninety years later, promoted an identical view when he wrote of 'that man who owns or occupies an acre of land, and does not make it produce all it is capable of producing, is an enemy to his country'.[28] There were practical reasons for this belief that the pursuit of agricultural improvement was an act of patriotism. In addition to threatened food scarcities at the time of the Napoleonic and Crimean wars, the unprecedented growth of the nation's population fuelled anxieties over national food supplies in the long-term. Only with the growth of overseas trade in the second half of the nineteenth century was this spectre of peacetime scarcity effectively laid to rest.[29]

However, the promotion of agricultural improvement also embodied values of a more abstract kind than those associated with a practical patriotism. Improved agriculture was upheld as the very hallmark of Western civilisation. The report of a select committee appointed in 1795 to consider the means of promoting enclosure makes the point vividly:

> The idea of having Lands in Common, it has been justly remarked, is to be derived from that barbarous State of Society, when Men were Strangers to any higher Occupation than those of Hunters or Shepherds, or had only just tasted the Advantages to be reaped from the Cultivation of the Earth.[30]

Here it is made plain that the advance of (privately owned) permanent arable was considered to be, not simply an alternative farming system, but an advance in the scale of civilisation. J. C. Loudon envisaged a similar scale of farming in his *Encyclopaedia of Agriculture* (1835), when he placed open-field farming just above the 'economy of savages' in a category called 'barbarian agriculture'. Above these in the scale came 'the agriculture of habit' whilst 'the agriculture of science' represented the pinnacle of achievement.[31]

The agricultural improver was committed to arable farming, and this commitment was given up with difficulty and regret when the economic climate made arable systems less viable at the end of the nineteenth century. John Wrightson, for example, commenting on the increase in grassland and the decline of cultivation in 1888, said: 'it is a rather sad alternative in a scientific age like this'. Henry J. Webb similarly registered the unease produced by the extension of permanent pasture in his remark that 'it is a retrograde movement; that in an age of science and free education it is humiliating to have to give up tillage and return to primitive pasture'.[32] The late nineteenth century stood at the end of a period of over a century in which the message of agricultural improvement literature was unequivocally bound up with the expansion of arable and the reclamation of marginal land. Arthur Young wrote in the 1770s of 'grieving' at the sight of uncultivated tracts of 'improvable land' and endeavoured to prove that the 'blackest mountain' was capable of improvement. Nearly ninety years on,

J. L. Morton was urging the reclamation of 'thousands of acres lying in comparatively unprofitable pasture' in much the same fashion.[33]

The landscape change envisaged by agricultural improvers did not only involve the extension of the margins of cultivation into areas formerly kept as moorland, common, rough pasture or marsh. Also very much part of the improver's agenda was the re-organisation of existing farmland, particularly the enlargement of fields and the reduction of boundary hedges. This is graphically illustrated by Loudon in his comparison of a 350 acre farm in Middlesex 'before' and 'after' improvement. The original boundaries, consisting of copse wood, composed of hazel, dogwood, black- and white-thorns, wild roses, brambles and native shrubs, were removed or thinned to yield an additional 50 acres of cultivable land, whilst the number of fields was reduced from twenty-two to seventeen.[34]

The new farmsteads pictured by nineteenth-century agricultural improvers conjure up images of industrial production. D. G. F. Macdonald held out the farm of the Scottish Lothians as the type to which English farm design should aspire, where 'each farm-steading has its steam engine, tall chimney, and perfectly-arranged cluster of solidly built' outbuildings.[35] J. L. Morton thought that the farm was like a 'great machine, a manufactory of beef, milk, and other marketable commodities', and that the farm should be redesigned accordingly,[36] whilst J. J. Mechi's vision of the future was of 'a railway activity pervading agriculture, farms squared, trees removed, game moderated, [and] tramways intersecting estates'.[37]

The Agricultural Topography of South-West England

In many respects the agricultural topography of South-West England conflicted with the ideal topographies envisaged by the agricultural improvers. It was true that the landscape was predominantly enclosed, as agriculturalists approved, but the enclosures of many districts were small, the land-ownership and land-occupation was often inter-mixed, whilst acres of hedges and unimproved waste land seemed ample demonstration of a lack of improving zeal amongst the region's inhabitants. John Grant, a surveyor and land agent, commented that:

> Every practical farmer coming into Devonshire for the first time is struck with the fertility of the soil and the genial climate with which the county is favoured. He may not be equally struck with the quality of the farming . . . [and] he is certainly astonished at the small quantity of produce returned per acre, a much greater [amount] being produced in districts immeasurably behind Devon in natural advantages.[38]

One of the main reasons for the poor performance of the region's agriculture advanced by Grant was a topographical one, what he described as 'the baneful effect of the high hedgerows and small enclosures'. Grant estimated that up to 7 per cent of the land surface was occupied by hedgerows, and he argued that it was necessary to make extensive changes to this by removing field boundaries and consolidating fields. He backed up his case using the newly available data of the tithe apportionments from which topography, including field sizes, could be described statistically for the first time for a wide area. The result of his survey of ten parishes in the Exe Vale was to establish that 'there are 1,651 miles of hedge; about half as long again as the famous wall of China; or sufficient to hedge round the whole of England with an immense bank of earth'.[39] In addition to the obviously large physical acreage taken up by hedgerows, the agricultural improver objected to them because they provided a habitat for birds and 'vermin', shaded the crops, provided nurseries for weeds, impeded drainage, and cost a great deal to keep in proper repair. The remedy suggested was to follow the example of Lord Poltimore's steward, who had taken down eight miles of hedge on a 200 acre farm, thereby adding 15 acres to agricultural production.[40] Grant not only justified his arguments for radical change on the grounds of economics and common sense, he also added a sprinkle of divine inspiration in his comment that 'I trust [we] will not cease till our land has attained that full amount of productiveness which the Almighty intended, and which his kind providence has placed within reach of every practically scientific agriculturalist'.[41] Twenty years later, writing in 1865, D. G. F. Macdonald found the local agricultural topography as inimical to increased production as ever. Describing some Devon fields as being 'ball-room size', he contrasted the locality with Norfolk:

> In Norfolk, for example, there are farms of more than 1,000 acres divided into four portions only; whilst in Devonshire there are many farms which can scarcely find a tenant, because, though consisting of but 100 acres, they contain probably fifty fields. Thus the unfortunate tenant has fifty gates to open, he has got fifty headlands and fifty hedges full of trees. Can such a tenant live and prosper?[42]

Similar comments were made about Somerset. William Sturge commented that, 'I can have no hesitation in stating that it would be well, if possible, to sweep away one half of the fences throughout the greater portion of the county'.[43]

Not only were fields too small in the eyes of the agricultural improver, but the overall size of the average farm was considered to be too small for modern husbandry. In Cornwall, Karkeek advocated amalgamation because 'good husbandry' was in his view extremely rare on small farms.[44] In Devon, assistant tithe commissioners took an equally dim view of the agricultural capabilities of parishes dominated by small farms. Of East Buckland it was said that due to the small size of farms and a corresponding lack of capital, it was 'doomed not to participate in the improvement which [had] taken place during the past 25 years in the Midland counties of England'.[45]

Another cause of astonishment to visiting agricultural topographers was the vast extent of unimproved moorland. Charles Vancouver, surveying the scene in north Devon, thought that the extent of the 'wastes' could be best imagined by supposing the district submerged under water, the cultivated enclosures would then appear like a 'group of small islands in the middle of an ocean'.[46] Mid-century estimates of the extent of wasteland in the county of Devon alone put the figure at over 454,000 acres.[47] It was the fervently held belief of agricultural improvers that much of this land could be rendered productive by thorough drainage and reclamation.

Exmoor, however, posed a greater challenge to contemporary agricultural aspirations. The written descriptions of the assistant tithe commissioners are replete with images of a bleak, impassable, inaccessible, chilling and difficult environment. At High Bray, James Jerwood was convinced that the difficulties encountered in the natural environment here were sufficient to forestall any 'visionary scheme' to bring it into cultivation. Jerwood entertained no illusions of it ever being cultivated. At Challacombe, Jerwood spoke vividly of:

> a sort of *open* country where not a single tree hinders the wind from blowing upon you—or solitary bush to intercept the sunshine; their fences are formed of stone but nothing presumes to grow on them; there is plenty of room—you may ride for miles over the common—your road a sheep's track, and chance your only guide to find your way out of it.[48]

Robert Page's descriptions of the moorland parishes across the border in Somerset similarly spoke of inaccessibility, of bleak and exposed situations, and of unproductive land.[49]

Natural obstacles to improved agriculture were one thing, but the topography of South-West England was also marked by distinctive ways of farming the environment which the visiting agriculturalists had not encountered before. These practices involved the breaking up of pasture for temporary cultivation to produce two or three crops of grain. After this the land was laid down to pasture to remain a grass ley for several years before being broken up again. Convertible husbandry, as practised in north Devon in the early nineteenth century, meant that only an estimated 13 per cent of the land was either under corn cultivation, or in preparation for corn cultivation, at any one time. Of the pasture, 87 per cent was temporary grass subject to the convertible system, whilst only 13 per cent was permanent pasture, marsh and meadow.[50]

The operation of this system, also prevalent in Cornwall and parts of Somerset, meant that the landscape bore the marks of prior cultivation in even the remotest localities, whilst the land-use categories 'arable', 'pasture' and 'common' were blurred. As James Jerwood noted rather irritably in surveying the parish of St Giles-on-the-Heath: 'A great deal of the land in this parish which is termed arable is tilled only once in fifteen or twenty years. All this is only a trifle better than the common land—and ought to have been classed with it'. Frederick Leigh likewise found it impossible to separate arable and pasture in his valuation of the parish of Stoke Rivers.[51]

Associated with the convertible system were a battery of techniques and tools which were distinctive to the region. When William

Marshall wrote his survey of the region's agriculture in 1796 he listed twenty-seven items of agricultural practice which he believed marked the uniqueness of the peninsula compared with the rest of the country. Indeed, Marshall was convinced that he was beholding mysterious customs of great antiquity, associated with the first wave of foreign settlement in England. He coined the term 'Danmonian husbandry' to describe the practices he saw, after the ancient name for this western part of Britain, and argued that it could be observed in 'almost pristine purity' in the Tamar valley.[52]

In the manner of an anthropologist, Marshall believed he could trace the ancient origin of agricultural practices in the West of England to France. The practice of paring and burning pasture and moorland for cultivation he names 'Devonshiring', from an 'old tract' he had seen some years previously in the British Museum. Again he ascribed the practice to 'a time beyond which [neither] memory nor tradition reaches', and concluded that it was imported from the Continent. He viewed the practice of making water-meadows in the same light.[53]

Not all the agricultural topographers experienced the degree of 'culture shock' which Marshall expressed in his writings on the region. Nevertheless, a considerable strand of incomprehension continued to lace agricultural accounts of the South West in the following century. Measured against the agricultural systems of Midland and Eastern England, it seemed to all that radical change in the farming practices of the South West was essential. Extensive wastes, 'shifting cultivation', small farms, and tiny enclosures flanked by enormous hedgebanks, were just some of the topographical elements which came in for criticism under the scientific gaze.

The Modernisation of Agricultural Landscapes in the South West, *c.*1790–1875

Although agricultural writers all tended to emphasise what was unique about the agricultural systems and topography of the South West, giving something of a timeless quality to their descriptions, it is nevertheless clear from the information they give that the region was undergoing considerable change during this period. Of all the agricultural landscapes of the South West, that of Dorset underwent the earliest and most rapid change. In 1793 John Claridge estimated that only a third of the county's land was under tillage, and remarked that the 'open and uninclosed parts, covered by numerous flocks of sheep, scattered over the Downs' was the most striking feature of the county's landscape.[54] By the 1850s this landscape of rough and coarse pastures was being replaced by swedes, wheat, barley and clover. In 1854 L. H. Ruegg reported that between Dorchester and Blandford there was scarcely a parish remaining in which the downs had not been broken up. By this date thousands of acres had already been converted to arable, a practice which was increasing with each successive year.[55] By the 1880s it was admitted that the conversion of pasture into arable had been pushed too far. W. C. Little commented that the downs had been broken up at considerable elevations and many would have been better left 'maiden down'.[56]

Traditional systems of farming, which included extensive sheep-grazing on permanent pasture, and rotations of three corn crops in succession followed by a fallow without the intervention of green crops or artificial grasses, were dying out. In their place came the 'Norfolk system' of four-course or five-course rotations, incorporating wheat, turnips, barley and clover. Sheep were increasingly fed on green crops and artificial grasses, and the water-meadows, a distinctive feature of traditional sheep-farming systems, were rendered less necessary by these introductions.[57]

In Dorset as a whole L. H. Ruegg estimated that 12,000–14,000 acres of waste had been enclosed before 1800, and a further 53,000 acres between 1800 and 1854. One of the areas which was dramatically altered following enclosure was the Cranborne Chase district. Formerly a 'free warren', mainly consisting of hazel wood, the Chase had been home to an estimated 12,000 deer which roamed more or less freely over the area. The years following the passage of the Act to enclose the Chase in 1828 saw the extinction of the deer, and the breaking-up of an estimated 4,000 acres of downs, commons, and coppices.[58]

Accompanying these changes in the agriculture of the chalk districts, old life-leasehold tenures were gradually extinguished and farms were consolidated, a practice so extensive that Stevenson speculated in 1812 that the county's population growth had been slowed by it.[59] Farms of 1,000–2,000 acres were not unknown by the early nineteenth century. When government statistics on the size of farm holdings became available, the contrast between Dorset and other counties of the South West was fully revealed. For example, in 1880 only 15 per cent of Dorset's land was farmed in units under 100 acres in size, 28 per cent comprised holdings of between 500 and 1,000 acres, and as much as 15 per cent of the land was farmed in units over 1,000 acres. Comparable figures for Devon were 32 per cent, 2 per cent and 0.08 per cent respectively.[60]

Changes in the agricultural landscapes of the clay vales and the heathlands of Dorset proceeded more slowly. In the grazing and dairying district of the Vale of Blackmore small farms continued to operate and, as late as 1854, Ruegg noted that most were held on lifehold tenures. It was this fact that Ruegg used to explain the 'multiplication of fences' and the neglect of underdrainage in the district. The heaths remained even more resistant to change. Despite nearly ninety years of exhortations from agricultural improvers to reclaim or afforest the heathlands, most remained 'wretchedly barren'. The age of agricultural expansion had nevertheless left its mark on these areas and tended to impoverish them further. Ruegg noted that the heaths had been 'robbed outrageously', 'and their remorseless plunderers, after robbing them of all their possessions, *steal their skins*', a reference to furze and turf-cutting.[61]

In Devon and Cornwall traditional agricultural systems and the landscape associated with them altered more slowly; most changes were confined to the period after 1850. In mid-nineteenth century Devon, the land under tillage was still managed predominantly under the 'old Devon course' of turnips, wheat, barley, and oats, followed by a grass ley for two to six years. The system was beginning to be modified or replaced in the neighbourhood of large towns, and in the red sandstone districts. In areas like the Exe Vale, a four-course or five-course rotation of turnips, barley or oats, clover or grass for one or two years, followed by wheat, was now common practice. In the rest of the county the general neglect of improvements to tillage was attributed to the overwhelming importance of cattle and sheep in the farm economy.[62]

Field patterns and hedge boundaries changed relatively little in these circumstances. Although the value of hedge timber as fuel had largely diminished, the Devonshire farmer was still motivated to preserve small enclosures by the experience that small pastures supported more cattle than an equal acreage in large fields, and afforded more shelter for the stock.[63] H. S. A. Fox has added much to our understanding of the interrelationship between the field patterns of Devon and the traditional farming practices of the region. He points out that small closes were adaptations to convertible husbandry, as represented in the 'old Devon course' outlined above. This rotation cycle was between six to ten years long, rather shorter than was common in many parts of Devon and Cornwall during the preceding centuries, but nevertheless requiring a farm to be divided into a minimum of six to ten separate closes. On a small farm this would necessarily dictate small field sizes.[64] Variation in soil quality was also given as a reason for the retention of small enclosures. In parts of Somerset, for example, Acland noted that the texture of the land varied from field to field.[65]

Where new rotations were being introduced, and the rotation cycle shortened to four to five years, fewer closes were needed. Additionally, the gradual replacement of traditional hand-tool technology with horse-drawn machinery made the old closes increasingly inconvenient and uneconomic. It is no surprise to find that reports of active consolidation of farm boundaries were largely drawn from the Exe Vale where these changes were occurring most rapidly. By the mid-nineteenth century improving landlords in this area were expending considerable effort in enlarging enclosures and straightening boundaries. However, in 1856 the Exeter land-surveyor Robert Dymond estimated that the average size of fields was little more than three to four acres, despite this process of consolidation.[66] Although the process of enlargement continued in the second half of the century, Devon's small fields were still attracting adverse comment into the twentieth century.[67]

Other changes in the Devon landscape during the first three-quarters of the nineteenth century included the extension of the margins of cultivation and improved grassland into areas of heathland and 'mountain pasture'. Already by the time of the tithe surveys most of the lowland heath in the Vale of Exeter had disappeared. However, tracts of wasteland in the Woodbury and Haldon areas remained in 1849, together with extensive tracts of moorlands in Dartmoor and on the borders of Exmoor. Attempts at reclamation were made in these areas and seventy enclosure awards affecting a total of 51,000 acres were passed between 1800 and 1869.[68] On Exmoor, the Knight family abandoned their vision of arable farming, which had seen 2,500 acres pared, limed and ploughed in the early part of the century. However, the systems of improved stock and sheep farming which were subsequently introduced there meant that almost 10,000 acres of natural grass pasture were ploughed after 1866.[69]

The agricultural landscape of Cornwall reflected similar changes in farming practices as traditional convertible husbandry was modified to incorporate turnips and green crops, and wasteland was gradually reclaimed. In 1795 it was estimated that wasteland covered 500,000 acres, whereas W. F. Karkeek's estimate of 1845 was 200,000 acres.[70] A comparison of data on land-use in *c.*1836 and 1875 suggests that in Devon and Cornwall as a whole arable land had increased from 580,000 acres to 700,000 acres, reflecting a massive expansion in the acreage under oats, roots and green crops, at the expense of grassland.[71] In the second half of the century the development of the railway network stimulated further changes, including the growth of specialised market-gardening and dairying for distant markets.[72] Cornwall continued to be a county of small farms; by the end of the period less than 10 per cent of the agricultural area was farmed in holdings larger than 300 acres. This undoubtedly reflected the long-term influence of mining on the rural economy which stimulated demand for small-holdings, and was a factor militating against the modernisation of farm and field-layout so dear to the heart of the agricultural improver.[73]

This period was a significant one in the history of the Somerset landscape too. The old-enclosed landscape was subject to the same forces of inertia affecting Devon and Cornwall, namely the 'smallness and inter-mixture of the estates', particularly in the eastern district. William Sturge complained in 1851 that small enclosures, with crooked and ill-arranged fences, still remained the predominant feature of the Somerset landscape.[74] However, on the unenclosed moorlands, commons and wastelands the pace of change was quickening. Between 1800 and 1873, 116 enclosure awards were made, affecting 84,000 acres.[75] Reclamation and the introduction of more intensive agricultural systems were features of both upland and lowland areas. In the wetlands of the Somerset Levels a growing network of drainage channels and rhynes established new patterns of fields, roads and settlements, although failing to eradicate altogether the problems of seasonal flooding.[76]

Agricultural Topographers and the Visual Pleasures of Landscape

Agricultural topographers evaluated landscape in ways that were not entirely concerned with questions of agricultural potential. They also responded to the beauty of the region's landscape in a manner that reflected the artistic and literary conventions of the age. It would be misleading to focus on the purely agronomic aspects of texts on agricultural improvement in isolation from these developments. The idea of 'improvement', from its inception, had always embodied utilitarian *and* aesthetic aspects, and individual topographers often combined careers in estate management and landscape design: William Marshall and J. C. Loudon, for example, wrote manuals on both topics.[77]

Viewing landscapes was a popular pastime amongst the middle and upper classes of eighteenth and nineteenth-century England. The greater ease of travel, and the acquaintance of more people with the landscapes of foreign countries included on the 'Grand Tour', in addition to a growing internal tourist industry, added a further stimulus to looking at landscape afresh. So popular indeed was this pastime, that it has been remarked that 'it was impossible in the eighteenth century to avoid having views on landscape'.[78] It was during this intellectual ferment that the West of England became an established tourist destination, another development which influenced the perceptions of agricultural topographers. The first wave of visitors came in search of the health-giving properties of the mild climate and the sea. Robert Fraser notes in his *General View* of Devon in 1794 (Figures 7.3 and 7.4) that medical men were recommending patients to visit south Devon and Cornwall in preference to Lisbon.[79] In 1793 John Claridge noted that for twenty years there had been an annual influx of visitors to Lyme and Weymouth in Dorset, coming to enjoy the bathing. An interest in natural history also drew visitors to the coast in search of sea-shells, seaweed and fossils.[80] By the end of the eighteenth century tourists were coming to the region for the additional attractions of scenery.[81]

GENERAL VIEW

OF THE

COUNTY OF DEVON.

WITH

OBSERVATIONS ON THE MEANS OF ITS IMPROVEMENT.

BY ROBERT FRASER, A. M.

DRAWN UP FOR THE CONSIDERATION OF THE BOARD OF AGRICULTURE
AND INTERNAL IMPROVEMENT.

LONDON:

PRINTED BY C. MACRAE.

M DCC XCIV.

Figure 7.3 Title Page of Robert Fraser's *General View of the Agriculture of Devon*, 1794.

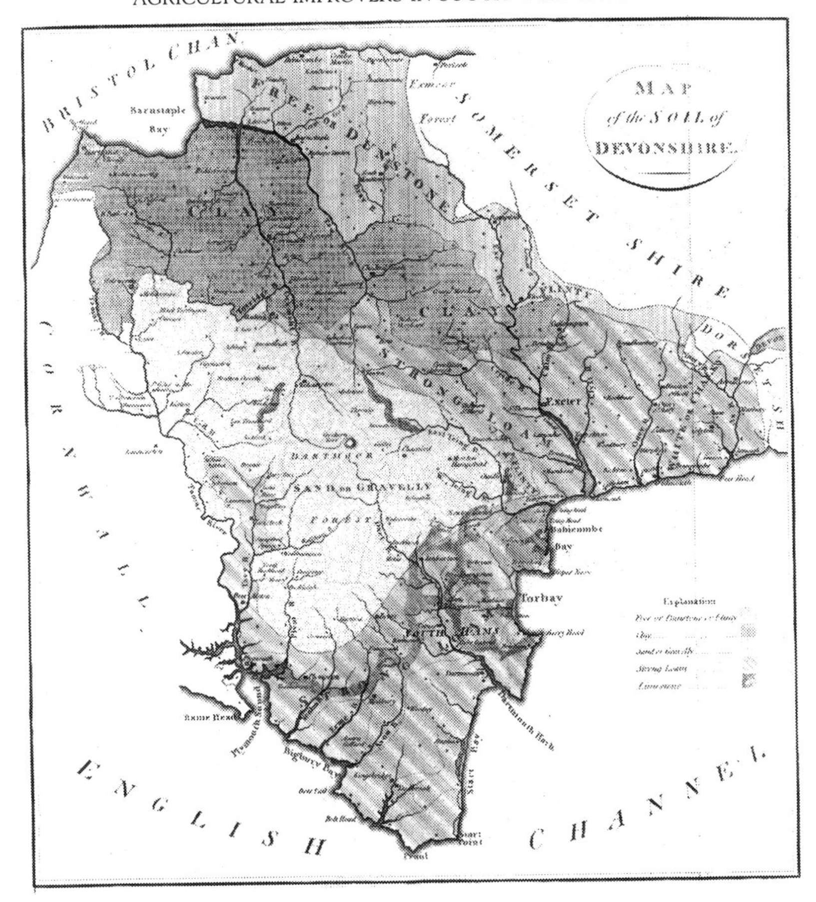

Figure 7.4 Map of the Soil of Devonshire from Robert Fraser's *General View of the Agriculture of Devon*, 1794.

Developments during the nineteenth century encompassed not only the growth of mass tourism centred on coastal resorts, but also a steady growth in the inland leisure industry. For example, Rider Haggard notes that Minehead owed its prosperity as a 'watering-place' in the early 1900s to the popularity of stag-hunting on neighbouring Exmoor, which brought 'numbers of rich people from London and elsewhere to share the sport'.[82] However, the key change in the appreciation of scenery to which the West of England owed its reputation as a tourist destination occurred at the end of the eighteenth century.

Whilst in the early eighteenth century the beauty of landscape had been equated with notions of fertility and usefulness to man, the second half of the century saw the ascendancy of an increasing taste for the wild and uncultivated elements of the landscape, in addition to an increasing naturalism in the design of landscape gardens.[83] In the middle of the eighteenth century two theories of aesthetics in nature gained popularity: the picturesque and the sublime. The picturesque was a term coined by William Gilpin in 1768 meaning 'that particular kind of beauty which is agreeable in a picture'. The sublime landscape on the other hand was defined by Edmund Burke in 1756 as an obscure, vast, difficult landscape, giving rise to feelings of 'delightful horror, a sort of tranquillity tinged with terror'. Landscapes suggesting infinity and vastness were sublime, including mountains, waterfalls and stormy skies.[84] What these aesthetic theories shared in common was an increasing opposition to the tamed, rationalised, and modernised landscapes promoted by agricultural improvers. William Gilpin, one of the important arbiters of picturesque taste, wrote that: 'Wherever man appears with his tools, deformity follows his steps. His spade and his plough, his hedge and his furrow, make shocking encroachments on the simplicity and elegance of landscape'.[85]

To Gilpin the regularity of corn fields was ugly, a 'natural' landscape was more beautiful than a cultivated one. This distaste for cultivated landscapes was reflected in contemporary landscape painting; Hugh Prince has described the search for depictions of newly enclosed fields, new farmsteads and reclaimed heaths in eighteenth century art as 'fruitless'. Painters and their patrons preferred landscapes without the intrusion of agriculture.[86]

Despite a refusal on the part of aesthetic theorists to recognise any connections between their own attitudes to land and those of agricultural improvers, there were in fact many parallels. Firstly, the aesthetic approach to landscape was itself like the 'scientific gaze' in the sense that it imposed an abstract structure of rules defining what was 'beautiful', 'picturesque' or 'sublime' in the evaluation of regional landscapes, in very much the same way as the agricultural improver evaluated landscapes according to national ideals of agricultural improvement. In common with agricultural improvement, the aesthetics of landscape expressed a desire to manipulate land and improve 'nature'. Secondly, both agricultural and aesthetic improvement reflected the increasing practical capability of landowners to transform landscape. Thirdly, the ability of landscape topographers to make a living by selling descriptive tours of the countryside was itself symptomatic of increasing rural mobility and prosperity derived from the profits of improved farming and trade.[87]

The revolution in landscape taste occurring at the end of the eighteenth century is of importance in the interpretation of agricultural texts because it is clear that many agricultural topographers were themselves profoundly influenced by these new ways of looking at the land, although they continued to praise cultivated landscapes. For example, G. B. Worgan began his *General view* of Cornwall (1811) by acknowledging four ways of evaluating the landscape, of which only one was agricultural. The three other responses to the country which he defines are the picturesque, the sublime and the antiquarian. Picturesque landscape is described by Worgan as: ' [landscape] pleasingly broken into hill and dale; some of the valleys . . . beautifully picturesque, and richly diversified with corn, woods, coppices, orchards, running waters and verdant meadows'. The important elements of beauty in landscape expressed here are diversity and richness. Worgan discriminates sharply in his account between picturesque landscape such as this, and sublime landscape, of which the following is a description: '. . . stupendous rocks, which form great barriers against the ocean, particularly about Land's End and the Lizard, filling the mind with awe, and giving rise to those solemn and sublime sensations, which elevate the heart to an *Almighty, all-creating Power*'. The antiquarian admirer of landscape, on the other hand, is directed to the 'multifarious druidical and Roman remains of karns, rock-basons, cromlechs, circles, religious and military enclosures'.[88]

Worgan's approach to the topography of Cornwall is in fact a sophisticated response to landscape consciously mediated by the prevailing aesthetic theories of the day. He finds pleasure in the landscape parks, wooded valleys, coastlines, and panoramas, including fertile cultivated fields. The only areas which strike him as ugly are the mining districts and the barren moorlands, by this date an old-fashioned view and a point to which I will return later.[89]

There are many points of similarity in William Marshall's response to the scenery of the region, written fifteen years earlier. Delight in agricultural landscapes is expressed in his response to the richness and fertility of the many river valleys, including the Dart and the Exe. He shares with Worgan a particular delight in woodland scenery, particularly wooded valleys. Rapid 'torrents', and wild coppices hanging on the hillsides, were to him 'most romantic and picturesque', and he was inspired by scenes like the ruins of Okehampton Castle and the wooded valleys surrounding Torrington, Bideford, Modbury and Ivybridge.[90]

Marshall's eye was 'delighted with the wilder scenery of nature' and he goes a step beyond Worgan in his appreciation of the uncultivated elements of landscape in that he incorporates the moors into his descriptions of beauty. It is not just that the 'mountain heights of Dartmore' provide a frame and a contrast to the rich valley landscapes, although this is an important part of the beauty which they are felt to impart to the scene. Marshall clearly takes delight in the moors themselves:

> The SURFACE of Dartmore proper, is truly mountainous. The composition is grand . . . the summits of several of the higher swells of Dartmore are truly savage, and rendered finely picturesque, by reason of immense piles of stones, or huge fragments of rock, thrown confusedly together, in the most grotesque manner.[91]

Charles Vancouver's views on beauty in landscape are very close to Marshall's. He admires the diversity and fertility of agricultural scenes, praises the designed scenery of landscape parks, derives pleasure from the contrast between the cultivated and the wilder landscapes, and is deeply affected by romantic scenery. In this respect, like Worgan and Marshall, he is particularly fond of woodland scenery, as his graphic description of his descent to Clovelly Court or his approach to Hartland Abbey illustrates. His writing on these areas is extremely interesting for he sees them as 'picturesque', 'wild' *and* productive, which he says 'all conspire[s] to excite the most agreeable sensations of grandeur combined with fertility and plenty in the highest degree'.[92]

Vancouver is also extremely attracted to barren landscape. Like Marshall he admires moorland scenery, describing summits 'capped with broken and craggy rocks' as 'wild and astonishing'. To him

Dartmoor rises in 'bold majestic grandeur'. He also loves the landscape of Braunton Burrows, with its sand dunes forming 'bold fantastic shapes' and its 'wild and sterile aspect'. Northam Burrows he describes as a 'beautiful and extensive common'. Again the contrast between the cultivated and the wild enthrals him. The area has, he says, 'an air singularly picturesque, by combining with the view of a desart that of a verdant plain or blooming valley'.[93] Vancouver viewed wild and cultivated land as complementary in terms of aesthetics. There is no sense that agriculture *mars* the beauty of any scene. The Exe Vale, with its 'fruitful orchards and thriving woodlands' and its houses and 'improvements' is also for him 'beautiful country'.[94] There is no sense in these early nineteenth-century agricultural topographies that development ruins the landscape.

With the demise of the Board of Agriculture, this expansive style of writing on landscape in regional essays on agriculture disappeared. The reports of the assistant tithe commissioners, and the essays on agriculture subsequently sponsored by the Royal Agricultural Society after 1840, are much more business-like texts. They include little information on landscape beyond that which is strictly relevant to the business of farming and landowning. However, there were occasional lapses. The assistant tithe commissioners reporting on the parishes of Oare and Stoke Pero describe the landscape as 'wild and romantic'.[95] Thomas Dyke Acland waxed lyrical on the Quantock Hills, 'intersected by the beautiful valleys of the Seven Wells, and Cockercombe, the classic ground of Wordsworth and Coleridge in the early days'.[96] William Sturge was moved to write appreciatively of the landscape of Exmoor with its summer bloom of 'fragrant purple and yellow flowers on the heath and furze'.[97] But these are exceptional passages in all cases.

Thomas Dyke Acland was nevertheless important in promoting a view of landscape wider than that implied by the needs of agricultural production alone. Under Acland's auspices, the Bath and West Society set up an arts section which first exhibited at the Barnstaple show in 1860[98]. The arts section was particularly active between 1860 and 1880, aiming to raise public taste by borrowing the best contemporary works of art, and to provide a forum for local artists to exhibit, and perhaps sell their work. Acland's aims were loftier still for he saw fine art as having 'a message to deliver to the agriculturalist and to the manufacturer, as well as to men of letters and men of cultivated taste. A part of that message is to teach them to cherish the love of Nature.'[99] It is interesting that agricultural topographers expressed views on landscape which addressed wider questions of beauty, and it reinforces the point that agricultural texts were not simple descriptions but were shaped by wider cultural values. Given that the prevailing aesthetic culture increasingly admired the wilder side of nature, the agricultural topographer, with his professional interest in 'improving' land, was increasingly faced with a conflict of values.

The Changing Landscape: Agricultural Production Versus Beauty?

John Barrell has examined how agricultural improvers dealt with this problem of the agricultural landscape being increasingly viewed with dislike in polite society. Barrell argues that only those writers oblivious to the revolution taking place in landscape tastes were able to reconcile their practical with their aesthetic interests.[100] In a sense this is true; the two sets of values are deeply conflicting and cannot be resolved. However, the evidence of the agricultural topographies of the South West suggests that the visual pleasures of cultivated and fertile land could co-exist much more happily with the new taste for 'wild' nature than Barrell allows.

There was certainly an awareness in some agricultural writings of a conflict between the needs of agricultural production and the aesthetic qualities of landscape. This is particularly apparent in agricultural writing on hedges. For example, William Marshall discovered fields in the South Hams which approximated the ideal field lay-outs discussed earlier, but was *not* delighted when he encountered them. Indeed, he sounds rather disappointed with the landscape:

> Notwithstanding the extraordinary beauty of the ground, or natural surface, of this District, it is far from being rich in picturable scenery. Square fields, and straight lines of Hedgewood, how profitable soever they may be to the Farmer, and pleasurable to the mind reflecting on their utility, are not grateful to the eye, viewing them in the light of Ornament.[101]

J. C. Loudon's summary of current agricultural thinking on hedges in 1835 admitted that there was a dilemma. However, he believed that 'there can be no permanent beauty that is inconsistent with utility', and advised the keeping of hedges free from timber trees despite the 'garden-like appearance which they give to the landscape'. The effect of removing hedges altogether, in order to 'correct' the outlines of fields, could be softened by leaving single trees in place of the eradicated hedgerows.[102] Thomas Acland's report on Somerset included a critique of the number of hedges in the Vale of Taunton Dean, yet he also acknowledged that the beauty of the region 'depends, alas! in great measure on the richness and frequency of its hedgerows'.[103]

John Grant's advice on removing hedgerows, which was discussed earlier, was offered so that Devon 'may soon become famous for its superior farming as it is now for its mild climate and beautiful scenery'. But herein lay the conflict, how to enhance the one without destroying the other. Admitting that hedges added so much to the beauty of the county, Grant's answer was to appeal to landscape gardening: a landscape gardener would not have arranged the 'timber' in 'long straggling hedgerows', but 'in clumps, belts, woods, which would serve for shelter and give the best effect'.[104]

Philip Pusey, commenting on Grant's remark, thought that even the *beauty* of Devon and Somerset could be enhanced by removing a large part of the hedges in order to give fuller scope to the visual effect of the few remaining trees:

> A few of the most beautiful trees, especially oaks, which from the depth of their roots are less injurious than ash or elm, may be spared, and will have a more picturesque effect than long lines of undistinguished foliage. The undulating lines of the surface thus unmasked, afford often a graceful landscape, with swelling knolls hidden before, and on these knolls the farmer will not grudge a little ground for single trees or clumps planted in commanding situations.[105]

Pusey envisages the land, thus improved, assuming the character of an 'arable park'.

I have quoted this passage at length because it illustrates some wider points regarding picturesque landscape taste. It is clear that the move to increasing naturalism in landscape tastes, taking place from the second half of the eighteenth century, did not inevitably imply that 'nature' was accepted uncritically as the standard of beauty. This passage underlines the need to arrange landscape carefully before it forms a pleasing prospect. The use of descriptive terms like 'swelling knolls' and 'commanding situations' is consciously drawn from the picturesque painting tradition. Pusey refers to the beauty of landscape being *unmasked* by the intervention of the improver. Pusey's taste is clearly not for nature wild and unalloyed: the intervention of man is required before the landscape can assume its full aesthetic potential. Although Pusey's particular interpretation of beauty in landscape, with its reference to knolls and clumps, was decidedly out of fashion by the time he was writing, he nevertheless is representative of the spirit of the picturesque. The taste for a less-controlled landscape had already made itself felt within the Picturesque movement at the end of the eighteenth century, but it nevertheless remained a carefully managed approach to nature.[106]

If most agricultural improvers could maintain that a compromise between modernising fields and maintaining landscape beauty was attainable, no such possibility was evident regarding the reclamation of 'barren' moors. G. B. Worgan was unusually old-fashioned in describing the Cornish moors in terms of 'deformity'. It is clear that although he was prepared to acknowledge that moorland had its economic uses, he was actually repelled by the sight of them.[107] Here his aesthetic values were fully in harmony with his agricultural values. In contrast, Marshall, Vancouver and Sturge were able to include barren land in their conceptions of beautiful landscape. Clearly their aesthetic responses did not square at all with their fervently expressed desire, as agricultural improvers, to see these areas of land removed.

If there was, as Keith Thomas has argued, a sensibility in society which saw the productive use of land as ugly and distasteful by the early nineteenth century, the agriculturalists, unsurprisingly, did not really share it.[108] New complaints heard in some quarters against the spread of new buildings, roads, canals, and tourism were absent from agricultural texts which tended to see new development as enhancing the beauty of natural scenery. In Dorset, for example, Ruegg was delighted with the appearance of the tall chimney of the Revd Anthony Huxtable's experimental farm, in the midst of the downs. He argued that it lit up 'with sudden activity and animation a somewhat desolate district'.[109] However, a very real tension remained between the appreciation of unproductive landscape and the demands of modernised agriculture, even if the latter was given unquestioned priority.

Ultimately changes in the economy provided a resolution to this conflict. Under pressure from international competition, it was no longer economic to farm marginal lands in late nineteenth-century Britain. Land-use in the South-West region followed national trends: after 1875 the margin of cultivation retreated rapidly, and by 1900 the overall balance between arable and pasture was returning to that of the 1830s. By 1939 the steep hill-slopes, which had formerly carried crops of oats, had seen the return of bracken, gorse and brambles.[110]

This transition required a revolution in the way land was regarded. Few were as sanguine for the future as James Caird, who remarked in 1878 that:

> This country is becoming every ten years less and less of a farm, and more and more of a meadow, a garden, and a playground. The deer forest, and grouse, in the higher and wilder parts of the country, and the picturesque commons in the populous districts, are already, in many cases, *not only more attractive, but more remunerative in health and enjoyment, than they probably would be if subjected to costly improvement* by drainage, or by being broken up for cultivation.[111]

Such a view of land, abandoning as it does the scientific drive for maximum production, would have been incomprehensible to the agricultural improvers discussed here.

AID TO RESEARCH: A LIST OF HISTORICAL WORKS ON THE TOPOGRAPHY OF SOUTH-WEST COUNTIES

IAN MAXTED AND MARK BRAYSHAY

> In such a work as the present, entire completeness can scarcely be hoped for.
>
> (James Davidson, 1852)

This list has been compiled in order to help researchers interested in past topographical studies of Cornwall, Devon, Dorset and Somerset to identify the authors and their works. Inevitably, as Davidson noted when he compiled his bibliography of works on Devon, 'entire completeness' is virtually impossible and no claim is made that the following list is comprehensive. As work on this chapter proceeded, the Historical Manuscripts Commission announced that a survey of the papers of antiquarians was to be undertaken and it may therefore be noted that the outcome of that project may ultimately include further material of interest to South-West historians. It has, moreover, become clear that in order to cover all the available historical works on the topography of South-West England, an entire volume would be required. The compilation of a more complete catalogue might well in future be a worthwhile task for the Centre for South-Western Historical Studies.

In general, in this listing, studies focused on individual towns or districts in the South West and the travel writings of authors who offer very little useful topographical description have been deliberately omitted. Instead, we have decided to include only those key works which describe the topography either of an entire South-Western county, or the region as a whole, plus the work of the 'agricultural topographers' and some 'travel writers' where the topographical content is deemed to be particularly significant. While the majority of the material included here has been personally inspected by the compilers, there is also a significant number of items suggested for inclusion by librarians, archivists and others, which we have not ourselves looked at. Moreover, we have not inspected every copy of every work which we cite in the list.

Although a considerable amount of topographical writing was published in the nineteenth century, surprisingly few of the authors added genuinely new insights. In compiling this listing it was therefore decided that only a small sample of the better Victorian topographers would be included. However, attention is drawn to other key catalogues and guides to topographical and historical studies of the four South-West counties, and an attempt has been made wherever possible to indicate at least one library or other collection in the region which holds a copy of a work cited in the list. A number of the best studies progressed through several editions, sometimes issued in an altered or extended form, and this listing therefore attempts to cite works chronologically according to the date of the edition. In recent years, facsimile reprints have been published of some of the region's best-known topographical studies and, where relevant, these are also mentioned below. As information has been gleaned from a variety of sources, the descriptions given are not presented in a completely consistent fashion, but are as detailed as the information easily to hand would permit.

It is also important to note that many more manuscript sources exist on the topography of the South West than are listed below; these can be found not only in the six record offices located in Devon, Cornwall, Dorset and Somerset, but also in other county record offices throughout Britain and in national collections such as the British Library, the Bodleian Library, and the Public Record Office. In this checklist, manuscripts and published texts have been listed together. However, published texts can be readily distinguished as their titles are given in italics.

There are three sections, each arranged by county with a general section at the end:

1. The works of the major county historians. The arrangement of this section is chronological by historian within each county. Under each historian the arrangement is also chronological, as far as can be

ascertained for each compilation, and the manuscripts normally precede the printed editions.

2 Accounts of travel and agricultural surveys. The arrangement of this section is alphabetical by author within each county and the descriptions are generally shorter.

3. Bibliographies and guides to research. For the major county bibliographies some annotation is offered as a guide to the coverage.

In order to assist researchers in locating copies of the works cited in the list, wherever possible, in each case at least one South-West library collection is noted where a copy is available, in some instances giving (within parentheses) the shelf location within the collection. These locations are not exhaustive. For some items references to published catalogues or other literature on the work are also given.

Abbreviations

('?' in the listing denotes an estimated date)

BL	British Library
BRP	Barnstaple Public Library
CAM	Camborne School of Mines Library
CRO	Cornwall Record Office
DOR	Dorset County Library
DRO	Devon Record Office
DRR	Dorset Record Office
EXC	Exeter Cathedral Library
EXI	Devon and Exeter Institution
EXP	Exeter Public Library
EXU	University of Exeter Library
PEN	Penzance Public Library
PLP	Plymouth Public Library
PLU	University of Plymouth Library
RIC	Courtney Library & Cornish History Archive, Royal Institution of Cornwall
SAS	Somerset Archaeological & Natural History Society
SRO	Somerset Record Office
SSL	Somerset Studies Library, Taunton
TIV	Tiverton Library

TQP Torquay Public Library
WSL Westcountry Studies Library, Exeter
YEO Yeovil Public Library

SECTION 1: THE WORKS OF MAJOR COUNTY HISTORIANS

CORNWALL

1. CAREW, Richard
[Miscellaneous genealogical papers and correspondence]. – 1592–1615.
Locations: BL (Add. MSS 4421, 29300E, 34599, Ct Jul c iii 30b, Ct Jul f xi 261, App 67)

The survey of Cornwall. Written by Richard Carew, of Antonie, Esq. – London: Printed by S. S. for Iohn [i.e. John] Jaggard, and are to bee sold neere Temple-barre, at the signe of the Hand and Starre, 1602.
Refs: Upcott: Cornwall I; STC 4615
Printer: S. Stafford. Entered at Stationers' Hall 8 Feb
Locations: WSL, PLP

The survey of Cornwall. And an epistle concerning the excellencies of the English tongue. Now first published from the manuscript. By Richard Carew, of Antonie, Esq. With the life of the author, by H*** C***** Esq. – London: Printed for Samuel Chapman, at the Angel in Pall Mall; Daniel Brown, jun. at the Black Swan, without Temple-Bar; and James Woodman, at Cambden's Head, in Bow-street, Covent Garden, 1723. – [2], xix, [9]p, 159 leaves, [9], 13, [1]p; 4o.
Refs: Upcott: Cornwall II; ESTC t135304
H*** C*****=Pierre des Maiseaux

The survey of Cornwall. And an epistle concerning the excellencies of the English tongue. Now first published from the manuscript. By Richard Carew, of Antonie, Esq. With the life of the author, by H*** C***** Esq. – A new edition. – London: printed for B. Law; and J. Hewett, at Penzance, 1769. – xxv, [7], 159 leaves, [10], 13, [1]p; 8o.
Refs: ESTC t080564 (EXp)
List of subscribers (6 pages)
Locations: WSL (sCOR/1769/CAR), EXI (SW Cup AC 06 CAR), PLP

Carew's survey of Cornwall; to which is added, notes illustrative of its history and antiquities, by the late Thomas Tonkin, Esq. and now first published from the original manuscripts, by Francis Lord De Dunstanville. Likewise, a journal or minutes of the convocation or parliament of tinners for the stannaries of Cornwall, held at

Truro, in the year 1710. The grant of the sheriffallty to Edward, Duke of Cornwall etc. – London: Printed by T. Bensley, Bolt-court, Fleet-street, for J. Faulder, New Bond-street; and Rees and Curtis, Plymouth, 1811. – xxxix, 459p; 29cm.
Refs: Upcott: Cornwall III
Locations: WSL (sxCOR/1602/CAR), EXU (RC 942.37/CAR), EXC (S.37.CAR), PLP

The Survey of Cornwall, with an introduction by F. E. Halliday. – London: Melrose, 1953. – 334p.
Locations: EXU (942.37/CAR), CAM, PLP

The Survey of Cornwall, edited with an introduction by F. E. Halliday. – Facsimile edn – New York: A. M. Kelly, 1969. – 334p.
Locations: EXU (942.37/CAR), PLU

2. NORDEN, John
Speculi Britanniae pars: a topographical and historical description of Cornwall... – London: Printed by William Pearson, for the editor; and sold by Christopher Bateman, 1728. – [24], 104p, plates : ill, maps; 4o.
Refs: Upcott: Cornwall IV; ESTC t049601, t127847 (large paper issue)
Locations: EXI (SW Cup), PLP

J. Norden, *Speculi Britanniae Pars: A Topographical and Historical Description of Cornwall.* – Newcastle: Frank Graham, 1966.
Facsimile reprint of 1st edn published in 1728.
Locations: EXU (942.37/NOR/X), PLP

3. TONKIN, Thomas
MS volumes for a New Survey of the History and Antiquities of Cornwall (includes some surveys of particlar parishes, notes on antiquities, notes on the tin trade, etc.).
Locations: RIC (Tonkin A–G)

An Alphabetical Account of all the Parishes in Cornwall . . . completed to 1763, 3 vols.
Locations: RIC (Tonkin H–J)

The Natural History of Cornwall by Thomas Tonkin of Trevaunance, begun at Lambrigan Ano. Dom. 1700.
Locations: RIC (Tonkin K)

Copies of deeds, charters, grants and historical records.
Locations: RIC (Tonkin L)

Tonkin's 'Commonplace Book'.
Locations: RIC (Tonkin M)

Proposed History of Cornwall.
Locations: BL (Add. MS 27763).

4. HALS, William
Parochial History of Cornwall.
Locations: BL (Add. MS 29762)

[The compleat history of Cornwal]. – Truro: Andrew Brice, [1750?].
Refs: Upcott: Cornwall V; ESTC t031007
Title of blue wrapper to Number 1: Number 1 of the weekly publication of four full sheets (now printing at Truro) of the compleat history of Cornwal; general and parochial, written by William Hals. At foot: N.B. We begin publication with the second part of the work, . . . because . . . first part not yet completed.
Locations: PLP

5. BORLASE, William
For details of Borlase MSS see: Pool, P. *William Borlase*. – Truro: Royal Institution of Cornwall, 1986, Appendix, pp. 285–97.

Memorandums of excursions and expense, 1751–1758 (includes tours in Cornwall, 1753–1757)
Locations: RIC (Borlase S/41)

Colecteana Physica et Palaeographica, 1736 (includes notes on antiquarian and natural history).
Locations: RIC (Borlase A/7)

Copy of the Exeter Domesday Book relating to Cornwall, with notes by William Borlase (copy made 1752).
Locations: RIC (Borlase Q/20)

Addenda to the Antiquities and Natural History of Cornwall: Various notes, 1755–1762.
Locations: CRO (DD/EN 2002)

Parochial Memorandums of Cornwall, 1740 (topographical and archaeological material arranged by parishes).
Locations: BL (Egerton MS 2675), RIC (Microfilms)

Observations in Natural History, designed as an introduction to facilitate the Natural History of Cornwall.
Locations: PEN (Borlase H/12)

Tracts and Extracts (includes William Borlase's journal of his tour of the Isles of Scilly in 1752, some unpublished drawings, the draft of his paper on the Islands sent to the Royal Society in 1752, and other historical material).
Locations: PEN (Borlase V/42)

Topographical sketches, church monuments, epitaphs, etc., for the ecclesiastical and parochial history of Cornwall (c.1748).
Locations: DRO (Z19/16/1), RIC & CRO (Microfilms).

Original drawings and original MS for the Natural History of Cornwall.
Locations: Bodleian (Ashmolean MS 1823)

Original drawings of Cornish stone circles.
Locations: Bodleian (MS Top. Gen. B537)

Observations on the antiquities historical and monumental of the county of Cornwall. – Oxford: Printed by W. Jackson, in the High Street, 1754.
Refs: Upcott: Cornwall VI; ESTC t131270
Locations: EXU (RC 942.37/BOR/X), EXC (S37)

Antiquities historical and monumental of the county of Cornwall . . . with a vocabulary of the Cornu-British language. – 2nd edn – London: W. Bowyer and J. Nichols, 1769.
Refs: Upcott: Cornwall: VII; ESTC t139784
Locations: EXI, WSL

Antiquities historical and monumental of the county of Cornwall / W. Borlase; introduction by P. A. S. Pool and Charles Thomas. – Wakefield: E. P. Publishing, 1973. – xxii, xvi, 464p: ill, fold. map; 37cm.
Reprint of 1763 edn.
Locations: WSL (fCOR/0001/BOR), PLU, EXU

The natural history of Cornwall. – Oxford: Printed for the author by W. Jackson, 1758.
Refs: Upcott: Cornwall VIII; ESTC t139226
Locations: EXC, EXI, Morrab (author's copy with annotations for second edition)

The natural History of Cornwall. – Oxford, 1758. First edn, privately printed.
Locations: EXU

The natural History of Cornwall. – Wakefield, 1973. Reprint of 2nd edn (London, 1769), with a new introduction by P. A. S. Pool and Charles Thomas.
Locations: PLU

The natural History of Cornwall. – London: E. & W. Books, 1970.
Locations: CAM

6. POLWHELE, Richard

The history of Cornwall. By Richard Polwhele. – Falmouth: T. Flindell, printers, for Cadell and Davies, London, 1803. – 7 vols.
Refs: Upcott: Cornwall IX–XIII
Locations: EXU, PLP

The history of Cornwall. By Richard Polwhele. New edn, corrected and enlarged. – London : Printed for Law and Whitaker, 13, Ave-Maria-Lane, by Michell & Co., Truro, 1816.
Refs: Upcott: Cornwall IX–XIII
Locations: WSL

The history of Cornwall. By Richard Polwhele. – Dorking: Kohler and Coombes, 1978. – 3 vols, 48 leaves of plates: folded map; 28cm. – ISBN 0–903967–11–1. Facsimile reprint of original edn in 7 vols: Falmouth: T. Flindell, printers, for Cadell and Davies, London, 1803, with an introduction by A. L. Rowse.
Locations: WSL (xCOR/0001/POL), EXI, PLU, PLP

7. LYSONS, Samuel and Daniel

See also Devon section for fuller details of manuscripts.

[Correspondence relating to Cornwall]. – 1811–13.
Locations: CRO (NRA 27819: RO misc DDX94)

Topographical and historical account of the county of Cornwall. By the Rev. Daniel Lysons, . . . and Samuel Lysons, . . . – London: Printed for T. Cadell, Strand; and G. & A. Greenland, Poultry, [1814]. – cclii, 394p, plates: ill, map. Refs: Upcott: Cornwall XIV
Locations: WSL (sxCOR/0001/LYS. Acc: R89748. Presented by Mrs J. R. Powell, 1921. J. R. Powell's copy with MS annotations and typescript inserts. Lacks plates *1*, *2*, *5*, *9*, *10*, *21*, *23*, *24*, *25*, *30–36*, *38*), EXU, PLU, PLP.

8. GILBERT, Charles Sandoe

An historical survey of the county of Cornwall: to which is added a complete heraldry of the same; with numerous engravings. By C. S. Gilbert. – Plymouth-

Dock: J. Congdon, 1817–1820. – 3 vols, 34 leaves of plates: ill, map: 28cm.
Locations: WSL (sxCOR/0001/GIL), PLP

9. HITCHINS, Fortescue

The history of Cornwall, from the earliest records and traditions, to the present time. Compiled by Fortescue Hitchins and edited by Samuel Drew. – Helston: William Penaluna, 1824. – 2 vols; 26cm. Hitchens wrote no part of this work, as he died as he was about to commence it, yet the publishers retained his name on the title page. It was first published in parts between 1815 and 1817.
Locations: WSL (COR/0001/HIT)

10. GILBERT, Davies

The parochial history of Cornwall, founded on the manuscript histories of Mr. Hals and Mr. Tonkin; with additions and various appendices. By Davies Gilbert. – London: J. B. Nichols & Son, 1838. – 4 vols; 23cm.
Locations: WSL (sCOR/0001/GIL. Inv: 337877–80, 38371–4), EXI, PLP

11. PENALUNA, W.

An historical survey of the county of Cornwall etc. Compiled by the Printer. – Helston: W. Penaluna, 1838. – 2 vols: ill; 19cm.
Locations: WSL (sCOR/0001/PEN. Inv: B2750–1), PLP

An historical survey of the county of Cornwall etc. Compiled by the Printer. – 2nd edn – Helston: W. Penaluna, 1843. – *2 vols: ill; 19cm.*
Locations: WSL (sCOR/0001/PEN. Inv: B83017–7), PLP

12. POLSUE, Joseph

A complete parochial history of the county of Cornwall, compiled from the best authorities & corrected and improved upon from actual survey. By Joseph Polsue. – Truro : William Lake, 1867–72. – 4 vols; 25cm.
Locations: WSL (COR/0001/POL), EXU, EXI

Lake's Parochial History of the County of Cornwall. – Wakefield, 1974. 4 vols. Facsimile reprint with an introduction by Charles Thomas.
Locations: PLP

DEVON

1. HOOKER, John

A discourse of Devonshire and Cornwall, with blason of armes, etc., the

bishops of Exeter, the revenews of the deneries and parsonages and other gentlemen. – 1599. – 171p; 2o.
Dated on fo 51b: 1599, deleted and replaced by: 1600.
Extracts in *Trans. Dev. Assoc.*, 47 (1915) 334–48.
Locations: BL (Harleian MS 5827), WSL (Microfilm).

The synopsis chorographicall of Devonshire or an historical record of the province of Devon. – 1599. – 353p. A fair copy of the material in BL Harleian MS 5827.
Locations: DRO (Z19/18/9. Ex libris John Prince de Berry Pomeroy vicar: An: Do: 1686. Presented by Revd Charles T. Wickham of Winchester 1910, who acquired it from the family of Short of Bickham, Devon)

Hooker also compiled *The Description of the Citie of Excester*, edited by W. J. Harte, J. W. Schopp and H. Tapley Soper and published by the Devon and Cornwall Record Society, 1919–47, as well as several other works, published and unpublished on the City of Exeter.

2. WESTCOTE, Thomas
[A view of Devonshire]. – [1650?].
Locations: DRO (not examined)

[A view of Devonshire]. – [1650?].
Locations: DRO (Plymouth, not examined)

[A view of Devonshire]. – [1680?]. – 232p; 2o.
Locations: WSL (sxDEV/1630/WES. Inv: 110714 transferred from Devon Record Office)

A veiw [i.e. view] of Devonshire by Thomas Westcot gent. 1630. Transcribed by I. P. 1696. – 1696. – 2 vols; 2o.
Manuscript. Vol. 1: Introduction. Lib 1–5. – 266p. Vol. 2: Lib 5 chapter 6 – end pp. 267–374.
Locations: WSL (msfDEV/0001/WES Inv: 106085–6 Brooking Rowe bequest 1908)

A view of Devonshire in MDCXXX, with a pedigree of most of its gentry. By Thomas Westcote; edited by Revd George Oliver and Pitman Jones. – Exeter: William Roberts, 1845. – 649p; 22cm.
Refs: Upcott: Devon I
Locations: WSL (sDEV/1630/WES Inv: 160483 manuscript index to parishes bound in; sDEV/0001/WES Inv: 54411–2), EXI, PLP

'Index to personal names in Westcotes "View of Devonshire in 1630" and his Devonshire pedigrees, edition of 1845', by A. B. Prowse. *Trans. Dev. Assoc.*, 27 (1895), 443–85.

3. POLE, Sir William

Survey of Devon. – [1700?]. – Manuscript transcripts with notes from his collections.
Locations: Bodleian Library (MSS top Devon b 3–4; MS Phil-Rob c154)

Survey of Devon. – [1700?]. – Manuscript transcript and drawings of seals of English women.
Locations: BL (Add. MSS 5485, 28649)

Survey of Devon. – [1750?]. 2 vols.
Manuscript transcript of a copy taken by John Anstis and with notes by Sir Isaac Heard, Garter King of Arms. Brooking Rowe bequest.
Locations: DRO (Z19/18/13a–b)

Collections towards a description of the county of Devon. By Sir William Pole of Colcombe and Shute, Knt (who died A.D.1635). – London: Printed by J. Nichols; and sold by Messrs White and Son; Leigh and Sotheby; and Payne, 1791. – *[2], xviii, 568p; 4o.*
Refs: ESTC t182928
List of surviving Pole MSS see *Trans. Dev. Assoc.*, Printed from manuscript in hands of Pole's descendants.
Locations: WSL (sxDEV/0001/POL. Refs: 781467/38: J. R. Powell's copy. R45987 A. J. P. Skinner's copy, 12457 Brooking Rowe bequest 1908W), EXU, EXI, PLP, SAS

4. RISDON, Tristram

The decimes or a corographicall description of the county of Devon, with the citty & county of Exeter, conteyning matter of history, antiquity, chronology, the nature of the country the com[m]odityes & government thereof . . . collected by the vew & travell of Tristram Risdon of Winscott Gent. – [1630]. – 297p; 2o.
Manuscript. P.221 contains the date 1616.
Locations: WSL (sxDEV/0001/RIS. Inv: 110715. Transferred from Devon Record Office 1985. Purchased fom Commins, Ex libris T. N. Brushfield)

Geographicall description of the county of Devon, with the citty and county of Exon, containing history, antiquity, chronologie, gentry and gover[n]ment

thereof. Collected by T. Risdon of Winscott, Gent. – 1605 continued to ye yeare 1630. – [120]p; 20cm. – Manuscript, probably in Risdon's hand. Includes list of parishes with patrons and value.
Locations: WSL (DRO: D.3590. Inv:780606/11. Purchased from Stanley Crowe June 1978, cat. 88, item 152)

[The decimes or a chorographicall description of the county of Devon]. – [1650?].
Manuscript transcript or extracts by George Gregory (or Jeffery?).
Locations: DRO (Z19/18/8)

[The decimes or a chorographicall description of the county of Devon]. – [1650?].
Locations: WDRO (Plymouth unexamined)

[The decimes or a chorographicall description of the county of Devon]. – [1650?].
Locations: RIC (unexamined)

Notitia Devoniae or a giographical description of the countie of Devon with the citty and county of Exon . . . by Thomas Risdon gent de Winscott. – [1650?]. – 1 vol; 4o.
Manuscript. Also includes: Historical collections relating to ye counties of Cornwall & Devon. The antiquitie, ffoundation and building of the cathedrall church . . . by John Hoker [inverted at back:] A sermon preached . . . by Mr Gregory. Purchase date: 1658.
Locations: WSL (sDEV/0001/RIS. Inv: 4877 transferred from Devon Record Office 1987)

Geographicale description of the county of Devon with the city & county of Exon containeing history, antiquity chronology . . . collected by T: R: of Winscott, gent. – [1675?]. – 190 leaves; 2o.
Manuscript. Inscribed: J. Trelawny. Bookplate of Charles Bath.
Locations: WSL

A chorographicall description of Devonshire, with the city and county of Exceter. Containing matter of history, antiquity, chronology. . . . Collected by the travall of T: R: of Winscott, gent. – [1690?]. – 365, [16]p; 2o.
Manuscript. Annotations and index by James Davidson. Date evidence: pp. 2, 9.
Locations: WSL (msfDEV/0001/RIS. Inv: 9434 Kent Kingdon bequest 1908–9)

A chorographical description or decimes of the county of Devon; with the city and county of the city of Exeter, containing matters of history, antiquitie

and chronology . . . collected by the travail of T. R. of Winscott gent: – [1690?]. – 343, [18]p; 2o.
Manuscript. Pagination irregular. Binding stamped Iohn White 1697. Fly leaf inscribed Tho Taylor 1759.
Locations: WSL (msfDEV/0001/RIS. Inv: 43200)

A chorographicall description of Devonshire with the city and county of Exeter: containing matter of history, antiquity, chronology . . . collected by the travell of T: R of Winscott gentlemen. – [1700?]. – 240, [3]p; 2o.
Manuscript. Phillips ms 9067. Phillipps ms 9067.
Locations: WSL (msxDEV/0001/RIS Inv: 106066. Brooking Rowe bequest, 1908)

The chorographical description or, survey of the county of Devon, with the city and county of Exeter . . . collected by the travail of Tristram Risdon, of Winscott, gent. – London: Printed for E. Curll at the Dial and Bible against St. Dunstan's Church in Fleet Street, 1714. – xiv, 186, [6], 4p; 8o.
Refs: Upcott: Devon II; ESTC 117944
Locations: WSL (sDEV/1605/RIS. Inv: 57016 bound with: A continuation . . .), EXU, EXI, PLU, PLP

A *continuation of the survey of Devonshire*. By Tristram Risdon of Winscot, gent. – London Printed for E. Curll at the Dial an[d] Bible against St. Dunstan's Church in Fleet-Stre[et], 1714. – [8], 425p; 8o.
Refs: Upcott: Devon II; ESTC 117944
Includes material obtained from a manuscript lent by John Prince.
Locations: WSL (sDEV/1605/RIS. Inv: 57,016 bound with: *The chorographical description* . . .), EXC, EXH, PLP

The chorographical description, or, survey of the county of Devon, with the city and county of Exeter . . . collected by the travail of Tristram Risdon, of Winscott, gent. – London: Printed for W. Mears, and J. Hooke, 1723. – xiv, 186, [12]p; 8o.
Refs: ESTC t117943
Reissue of the 1714 edition with cancel title-page. Later reissue recorded 1723.
Locations: WSL (sDEV/1605/RIS. Inv: 108468 transferred from Devon Record Office)

A *continuation of the survey of Devonshire*. By Tristram Risdon of Winscot, gent. – London: Printed for W. Mears, and J. Hooke, 1723. – [2], 425, [5], 4p; 8o.
Refs: ESTC t117943, n044664 Includes list of subscribers. Reissue of the 1714 edn with cancel title-page. Later reissue recorded 1733.
Locations: WSL (sDEV/1605/RIS. Inv: 108469 transferred from Devon Record Office).

The chorographical description or Survey of the county of Devon. By Tristram Risdon. Printed from a genuine copy of the original manuscript; with considerable additions. – London: Printed for Rees and Curtis, Plymouth, 1811. – xvi, 442p; 23cm.
Refs: Upcott: Devon III
Uses manuscript of John Cole of Stonehouse.
Locations: WSL (sDEV/1605/RIS. Inv: 810217/39, 108470; sxDEV/0001/RIS. Acc: 12466. One of 50 copies printed in quarto on demy paper, Brooking Rowe bequest, 1908. Bound in at back: Index to personal names by A. B. Prowse. Typescript 'Risdon: "Survey of Devon". – index : copy of index by Davidson in E.C.L. copy no. B.-R. 12465'), EXC, EXU, EXI, PLU, PLP, SAS

The chorographical description or survey of the county of Devon. By Tristram Risdon. Printed from a genuine copy of the original manuscript; with considerable additions. – Barnstaple: Porcupines, 1970. – Reprint of edn of 1811.
Locations: WSL, EXI, EXU, BRP, PLP

The notebook of Tristram Risdon, 1608–1628, transcribed and edited from the original manuscript in the library of the Dean and Chapter of Exeter by J. Dallas and H.G. Porter (London, 1897).
Locations: WSL, PLP, SAS

Refs: J. M. Hawker, "Sketch of Risdon", *Trans. Dev. Assoc.*, 7 (1875), 79–83. Prowse, A. B. "Index to Risdon's survey of Devonshire: personal names: edition of 1811, including the additons to 1810", *Trans. Dev. Assoc.*, 26 (1894), 419–50.
Locations: WSL

5. CHAPPLE, William

[Volume with parallel paste-ups of Curll's edition of Risdon's *Chorographical description* and *Continuation* together with manuscript annotations and proofs of unpublished sections of Chapple's Review]. – *c*.1780.
Locations: WSL (deposited in DRO: 3590)

A review of part of Risdon's Survey of Devon; containing the general description of that county; with corrections, annotations and additions. By the late William Chapple, of Exeter. – Exeter: Printed and sold by R. Thorn in Fore Street. Sold also by T. Davies . . . and W. Shropshire . . . London; J. Fletcher, Oxford; and Messrs. Merrill, Cambridge, 1785. – [2]iv, vii, [1], 144p; 4o.
Includes list of subscribers.
Refs: Upcott: Devon IV; ESTC t069294

Locations: WSL (sDEV/0001/CHA Acc: 1159 MS inscription: Wm. Cotton FSA 1879, purchased by means of the Kent Kingdon bequest, typescript inserted in back: "A selective index by Prof. W. J. Harte."); WSL (sDEV/0001/CHA. Acc: 12467. James Davidson's copy with ms note: "This copy contains 8 leaves, from page 169 to 184, which are not to be found in any other. They were never published, and were given to me by Mr Penny, bookseller of Exeter who purchased them among some waste paper. See the Preface p. ii". Manuscript index by Davidson bound in at back. Brooking Rowe bequest, 1908 Bookplates and sewn sections destroyed when rebound); WSL (sDEV/0001/CHA. Inv: 12967), PLP, SAS

A review of part of Risdon's Survey of Devon, containing the general description of that county; with corrections, annotations and additions. By the late William Chapple. – Barnstaple: Porcupines, 1970. – viii, 144p; 25cm.
Reprint of 1785 edn.
Locations: WSL (DEV/0001/CHA), PLP

6. PRINCE, John
Danmonii orientales illustres: or, the worthies of Devon. A work, wherein the lives and fortunes of the most famous divines, statesmen, swordsmen, physicians, writers, and other eminent persons, natives of that most noble province . . . are memoriz'd, in an alphabetical order . . . by John Prince, vicar of Berry-Pomeroy, in the same county. – Exeter: Printed by Sam Farley, for Awnsham and John Churchill, at the Black Swan in Paternoster-Row, London; and Charles Yeo and Philip Bishop in Exon, 1701. – [18], 600p; 2o.
Refs: ESTC t140977
Locations: WSL (sf920.02/DEV/PRI. Acc: 781474/74, Ex libris A. W. Searley with MS index and bibliographical cuttings inserted), SAS.

Danmonii orientales illustres: or, the worthies of Devon. A work wherein the lives and fortunes of the most famous divines, statesmen, swordsmen, physicians, writers, and other eminent persons, natives of that most noble province . . . are memorized, in an alphabetical order . . . by John Prince, vicar of Berry-Pomeroy, in the same county. – A new edition, with notes. – London: Printed for Rees and Curtis, Plymouth; Edward Upham, Exeter; and Longman, Hurst, Rees and Orme, London, 1810. – xxxvii, [1], [785], [11]p, plates.
Locations: WSL (sx920.02/DEV/PRI. Acc: R7978, Inscribed: R. Bartlett 1860), SAS

[Danmonii orientales illustres. Volume 2]. – 1701–16.
Locations: WDRO (formerly in library of John Francis Gwyn at Ford Abbey. In 1846 purchased by Sir Thomas Phillips, later of Thirlestane House,

Cheltenham, acquired by WDRO after 1933. Refs: J. Brooking Rowe, "The second volume of John Prince's 'Worthies of Devon'" in *Trans. Dev. Assoc.*, 32 (1900), 307–8)

Danmonii orientales illustres or the wortheys of Devon: volume the second in which are memorized some scores of famous persons, as earls, barons, bishops and others who were natives of that noble province, not mentioned before, by John Prince, vicar of Berry Pomeroy in that county. . . . Finished anno Domini M.D.C.C.xvi. – 1846. – [6], 90, [3]p.
Locations: WSL (s920.02/DEV/PRI Acc: BR105893, manuscript of extracts transribed from the manuscript at Forde Abbey by James Davidson, May 1846. Bookplate of Davidson, bequeathed to Exeter City Library by J. Brooking Rowe, 1908. [also two other fragmentary transcripts in WSL according to Brockett])

7. POLWHELE, Richard

Historical views of Devonshire. In five volumes. Vol. 1. By Mr. Polwhele. – Exeter: printed by Trewman and Son, for Cadell, Dilly, and Murray, London, 1793. – xix, [1], 192, 191–214p; 4o. – Only vol. 1 was published.
Refs: ESTC t130755
Locations: WSL, EXU, EXI, PLP

The history of Devonshire. By Richard Polwhele. – 1793–1806. 3 vols.
Refs: Upcott: Devon VI; ESTC n017500
Locations: EXU, EXC, EXI, WSL, PLP, SAS

The history of Devonshire. By Richard Polwhele. – Dorking: Kohler and Coombes, 1977. – 3 vols: folded map; 36cm
Facsimile reprint of edition published: 1797–1806 with manuscript index by James Davidson.
Locations: EXU, EXI, PLP, PLU, BRP, TQP, WSL

8. LYSONS, Daniel and Samuel

Correspondence.
Letters from various persons in answer to queries on Devon.
Locations: BL (Add. MSS 9426–30), WSL (microfilm)

Notebook. – *c.*1807.
Statistical and general information on Devon.
Locations: BL (Add. MS 9449), WSL (microfilm)

Notebook. – *c.*1807–14.
Largely ecclesiastical with some notes on Devon markets, charters etc.
Locations: BL (Add. MS 9450), WSL (microfilm)

Topographical collections. – *c.*1809–18.
Topographical notes on Devon parishes in roughly alphabetical order.
Locations: BL (Add. MS 9464), WSL (microfilm)

Notebook. – *c.*1818.
Notes relating to Exeter.
Locations: BL (Add. MS 9467), WSL (microfilm)

Notebook. – *c.*1818.
Notes on north Devon churches.
Locations: BL (Add. MSS 9468), WSL (microfilm)

Refs: Sherlock, R.J. "Lyson's notes on Devonshire churches", *Devon & Cornwall Notes & Queries*, 26 (1955), 151–4. – Discusses the above manuscripts, amending the dates provided in the BL listings. There are many other MSS items in various collections including Yale University, John Rylands Library, West Sussex Record Office, Bodleian Library, Society of Antiquaries, although these do not necessarily relate to the South West. See also section on Cornwall.

Copies of letters received by Daniel and Samuel Lysons giving information concerning Devonshire parishes & families for inclusion in their Magna Britannia, vol 6, containing Devonshire. – [1850?]. – 1 vol; 30cm.
Manuscript. Gives references to BL Add. MSS.
Locations: WSL (sxDEV/0001/LYS. Ref: 56111. Transferred from Devon Record Office 1987).

Magna Britannia; being a concise topographical account of the several counties of Great Britain. Vol. 6: Devonshire. By Rev. Daniel Lyson and Samuel Lysons. – London: Thomas Cadell, 1822. – 2 vols, plates: ill; 35cm.
Locations: WSL (xDEV/0001/LYS; sxDEV/0001/LYS. Acc: 12459–61. 3 vol, extra-illustrated copy; sxDEV/0001/LYS Acc: 160521–2), EXU, EXC, EXI, PLU, WSL, PLP (including extra-illustrated set), BRP, SAS, TQP

9. MOORE, Thomas
The history of Devonshire, from the earliest period to the present, by the Revd Thomas Moore. Illustrated by a series of views, drawn & engraved by & under the direction of William Deeble. – London: Published by Robert Jennings, 62, Cheapside, 1829. [Engraved title, present in many copies of the large and small format editions. The plates were reissued by Virtue in 1833.]

This is a bibliographically complex work:
Wrapper title of large format edition: *The history and topography of the county*

of Devon, by the Revd Thomas Moore. Including outlines of the physical geography, geology, and natural history of the county, by E. W. Brayley, jun . . . – London: Robert Jennings. No. 1 [–15], [Sept 1] 1829 [–1830]; Jennings & Chaplin no. 16 [–25] 1830 [–1831].
Large format edn:
Vol. 1 (books 1–4): 400p (parts 1–25 1829–31)
Vol. 2 (book 1) Biography: 629, [3]p. Index (2p). Parts 26–48? 1831–33.
Locations: WSL (sxDEV/0001/MOO Acc: 12454–5; sxDEV/0001/MOO Parts in original wrappers: 1–24)

Small format edn (no wrappers traced for this):
Vol. 1 (books 1–4): 574p.
Vol. 2 (book 1): 908, [3]p. Index (3p).
Locations: WSL

It has not proved possible to ascertain whether copies in the following locations are large or small format editions:
Locations: EXC, EXI, PLP, PLU, TQP

The History of Devonshire, from the Earliest Period to the Present, Illustrated by a series of views, engraved by William Deeble, reprint (London, 1935).
Locations: EXC

10. STOCKDALE, Frederick Wilton Litchfield
A concise historical and topographical description of the county of Devon. – 1842. – 800p.
Refs: I. Stoyle, "F. W. L. Stockdale: begetter of the Stockdale Collection", *Devon Historian*, 46 (Apr 1993), 3–8.
Locations: EXI (Inner Lib., bay 51, cupboard. Presented with volumes of notes to the Exeter Diocesan Architectural Society by his son F. C. Stockdale, 1858)

11. WORTH, Richard Nicholls
A history of Devonshire. With sketches of its leading worthies. – London: Elliott Stock, 1886. – x, 347p; 26cm. – Reissued in 1886 as an "extra-illustrated" edition with 94 engraved plates. A "cheap edition" was issued in 1895.
Locations: WSL (sDEV/0001/WOR), BRP (B/0001/WOR), EXU (Crediton Library 1886/WOR & 942.35WOR), PLP, TQP

Dorset

1. GERARD, Thomas

John Coker, Survey of Dorset, 1622 (This is now known to be the work of Thomas Gerard)
Locations: DRR (D/BOC: Box 23)

A survey of Dorsetshire. Containing the antiquities and natural history of that county. . . . Publish'd from an original manuscript, written by the Reverend Mr. Coker, of Map[p]owder in the said county. – London: Printed for J. Wilcox, at the Green-Dragon, in Little-Britain, 1732. – [4], 128, [20]p, plates; 4o.
Refs: Upcott: Dorset I; ESTC t180788 (under John Coker)
Locations: DOR, DRO, EXI, SSL

Coker's survey of Dorsetshire. – Sherborne: Dorset Publishing Co., [1980]. – 175p. – ISBN 0–902129–20–1.
Reprint of edition of 1732. The work was first written by Thomas Gerard in the 1620s and was first published under the name of John Coker in 1732. This republication contains an Afterword by Rodney Legg.
Locations: DOR (Copy no. 101 of 500 copies), WSL (DOR/1625/GER Inv: 800731/03 Copy no 56 of 500 copies), YEO

2. HUTCHINS, John

[Dorset antiquarian collections and correspondence]
Locations: Bodleian Library (Gough MSS Dorset 1–36, Top gen d 2–3)

The history and antiquities of the county of Dorset, with a copy of the Domesday Book and the Inquisitio Gheldi for the county . . . By John Hutchins, M.A. – London: Printed by W. Bowyer and J. Nichols, 1774. – 2 vols, plates: ill, map; 2°.
Upcott: Dorset II; ESTC t099476
Selections published as: *A view of the principal towns . . . 1773* (ESTC t077345)
Locations: WSL (sfDOR/0001/HUT. Inv: 846308–9 transferred from Tiverton Library), DRO, EXI (S 58.3–4), DRR, DOR, SAS

The history and antiquities of the county of Dorset . . . By John Hutchins, M.A. – 2nd edn – London: Printed by John Nichols, 1796–1815. – 4 vols, plates: ill, map; 2°.
Refs: Upcott: Dorset III; ESTC t099475
Corrected, augmented and improved by R. Gough.
Locations: DRR, DRO, WSL, DOR (also set of "The annals and iconography of Dorsetshire and Dorset worthies" the large paper copy of the second edn of Hutchins from the Ashburton library, extra-illustrated, collected and arranged by A. M. Broadley, 13 vols with index to extra illustrations)

The history and antiquities of the county of Dorset. John Hutchins. – 3rd edn, by W. Shipp and J. W. Hodson – London: J. B. Nichols & Sons, 1861–74. – 4 vols, plates: ill, maps; 40cm.
Locations: WSL (sfDOR/0001/HUT), DRR, DOR, SAS, YEO

The history and antiquities of the county of Dorset. John Hutchins. – 3rd edn, by W. Shipp and J. W. Hodson; introduction by Robert Douch. – Wakefield: E.P. Publishing, 1973. – 4 vols, 135 leaves of plates: ill, maps; 36cm. – ISBN 0–85409–974–3. Reprint of edition originally published: London: J. B. Nichols & Sons, 1861–74.
Locations: WSL (fDOR/0001/HUT), DRO, EXU (942.33/HUT/X), EXI (AE 06 HUT/X), PLU, WSL, DOR

SOMERSET

1. GERARD, Thomas

The Particular Description of the County of Somerset. – [1633?].
Locations: Northamptonshire Record Office (FH 113)

The Particular Description of the County of Somerset, drawn up by Thomas Gerard of Trent, 1633; edited by E. H. Bates, – Taunton: Somerset Record Society, 1900. – (Somerset Record Society; vol. 15)
Locations: WSL (PER/SOM), EXI (SW periodicals), SAS, SSL, YEO

2. STRACHEY, John

History of Somerset. – c.1737.
Refs: M. McGarvie, "John Strachey, F.R.S. and the Antiquaries of Wessex in 1730", *Transactions of the Ancient Monuments Society*, 27 (1983), 77–104.
Locations: SRO (DD/SH 96–98)

3. CAREW, Thomas

Papers of the Hundred of Williton and free manors, parliamentary survey of Hundreds of Cannington, Andersfield and North Petherton etc. – c.1735–50.
Locations: SRO (DD/TB 19/2–3)

Papers relating to the history of Somersetshire, Manuscripts relating to the city of Bath, Papers relating to the Hundred of Wellow and several boroughs in the county etc. – c.1735–50.
Locations: SRO (DD//TB/Box 20/1–5)

Series of folders arranged by hundreds containing general notes on the hundred and individual parishes. – *c*.1740–50.
Locations: SRO (DD/TB/51/1)

4. PALMER, Thomas

Loose papers including notes on hundred of Chew and hundred of Williton. – [1770?].
Locations: SRO (DD/AH/21/1–2)

History of Somerset: introduction and Williton Hundred. – [1770?]. Near contemporary copy.
Locations: SRO (DD/AH/10)

History of Nettlecombe.
Locations: SRO (DD/WO/38/1/6)

5. COLLINSON, John

[MSS of the history and antiquities of Somerset]. – [1790?].
Locations: Bristol Record Office (NRA 1087 Smyth AC)

The history and antiquities of the county of Somerset, collected from authentick records, and an actual survey made by the late Mr *Edmund Rack, adorned with a map of the county, engravings of Roman and other reliques, town seals, baths, churches and gentlemen's seats*. – Bath: Printed by R. Cruttwell, 1791. – 3 vols. Refs: Upcott: Somerset II; ESTC n008218.
Locations: EXI (S 48.1–3), SRO (DD/SAS/HV68/1–6 grangerised set belonging to John De Havilland, bequeathed by Mr Edwin Sloper, 1905), SAS, SSL (J. A. Bennett's copy with MS annotations), YEO, Society of Antiquaries (grangerised by W. Adlam), BL (Add. MSS 38200–3, 33830, extra-illustrated by W. Phelps)

Index to Collinson's history of Somerset: including a supplemental index (alphabet and ordinary) to all the armorial bearings mentioned in the work. Edited by Revd F. W. Weaver, Revd E. M. Bates and J. R. Bramble. – Taunton: Barnicott and Pearce, 1898. – iv, 325p; 29cm.
Locations: WSL (sxSOM/0001/COL), SAS, SSL, YEO

The history and antiquities of the county of Somerset. By John Collinson; with a new introduction by Robert Dunning. – Gloucester: Alan Sutton, 1983. – ill, folded map; 35cm – ISBN 0–86299–033–3.
Reduced facsimile of edition published 1791 with index published 1898 and supplement published 1939.
Locations: WSL (fSOM/0001/COL), SSL, YEO

Proposals for Printing by Subscription the History of the County of Somerset, collected from authentic records, and an actual survey . . . The topographical and natural history by Mr Edmund Rack, secretary to the Agricultural and Philosophical Societies at Bath – Bath, 1874.

A Specimen of the History of the County of Somerset: being an account of the parishes of Chilcompton and Porlock. – Bath, 1785.

6. LOCKE, Richard

Survey of Somerset. 1795–1806.
Refs: F. M. Ward, *Supplement to Collinson's History of Somerset*. – Taunton: Barnicotts, 1939.
Locations: SRO (DD/SAS/PR/86), SAS, SSL, YEO

7. SAVAGE, James

History of Somerset. – [1830?]. – 2 vols. – Phillipps MSS 13436 (Williton & Kingsbury West Hundreds) & 13454 (Fairfield, North Curry and Andersfield Hundreds etc.).
Locations: SRO (DD/AH/11)

History of the Hundred of Carhampton, in the county of Somerset, from the best authorities. – Bristol: William Strong, 1830. – viii, xxiv, 662p; 22cm.
Locations: SAS, SSL (grangerised copy marked Phillipps MSS 13675)

8. PHELPS, William

History of parts of Somerset, consisting of interleaved sections of Collinson and loose papers. – c.1850.
Locations: SRO (DD/SAS/C/1193/24 – formerly R/18)

The history and antiquities of Somersetshire, being a general and parochial survey of that interesting county, to which is prefixed an historical introduction with a brief view of ecclesiastical history and an account of the druidical, Belgic-British, Roman,Saxon, Danish and Norman antiquities now extant. By Revd W. Phelps. – London: The Author, 1836. – *2 vols, plates: ill, maps; 29cm*.
Refs: M. McGarvie, "Phelps' county history", *Notes and Queries for Somerset and Dorset*, 32 (1986–90) 563–4; Keen, Laurence, "William Barnes and his wood-engravings for Phelps, History and Antiquities of Somerset", *Notes and Queries for Somerset and Dorset*, 32 (1986–90), 644–5.
Locations: WSL (xSOM/0001/PHE), EXI (AS 06 PHE/X), SAS, SSL, YEO

SECTION 2: ACCOUNTS OF TRAVEL AND AGRICULTURAL SURVEYS

CORNWALL

Cox, T. A *topographical, ecclesiastical and natural history of Cornwall, with pedigrees of all the noble families and gentry*. – London: E. Nutt for M. Nutt, 1720. – pp. 306–64. – From his *Magna Britannia et Hibernia antiqua et nova; or, a new survey of Great Britain*.
Locations: WSL (sCOR/1720/COX), PLP

Fraser, Robert. *General view of the County of Cornwall, with observations on the means of its improvement*. – London: C. Macrae, 1794. – 70p; 4o.
Locations: PLU, WSL (s630/COR/FRA)

Karkeek, William Floyd. 'On the farming of Cornwall', *Journal of the Royal Agricultural Society*, 6 (1845), 400–64.
Locations: WSL (s630/COR/CAR), PLP

Redding, Cyrus. An *illustrated itinerary of the county of Cornwall*. – London: How & Parsons, 1842. – 264p, 4 leaves of plates: ill; 27cm.
Locations: WSL (sCOR/1842/RED), EXI (S50.12), CAM (93 423.7 RED)

Staniforth, Thomas. *The Staniforth diary: a visit to Cornwall in 1800*, edited by Jean Hext. – 1965.
Locations: CRO (VL/9/6)

Stockdale, Frederick Wilton Litchfield. *Excursions in the county of Cornwall, comprising a concise historical and topographical delineation of the principal towns and villages*. – London: Simpkin & Marshall, 1824. – vii, 171, [7]p, plates: ill, map; 26cm
Locations: EXU (RC 942.37/STO), EXI (S52.7), CRO (VL/9/10), WSL (sCOR/1824/STO)

——*The Cornish tourist or excursions through Cornwall*. – 1834.
Locations: CRO (VL/9/13)

Swete, John. A Tour in Cornwall by J. Swete, 1780.
Locations: RIC (DJS/2/1–2)

Walis, John. *The Cornwall register; containing collections relative to the past & present state of the 209 parishes; forming the county, archdeaconry, parliamentary divisions and poor law unions of Cornwall. To which is added a brief view of the adjoining towns and parishes in Devon from Hartland to Plymouth*. – Bodmin: Liddell & Sons, 1847. – 476p; 19cm.

Locations: WSL (sCOR/0001/WAL), EXU (RC 942.37WAL), EXI (AC06WAL), BRP (C/1847/WAL)

Warner, Richard. *A Tour through Cornwall in the Autumn of 1808*. – Bath: Printed by R. Cruttwell, 1809. – 363p.
Locations: WSL (sCOR/1808/WAR), EXI (AC 02 WAR), CRO (VL/9/7)

Worgan, G. B. *General View of the Agriculture of the county of Cornwall, drawn up and published by the Board of Agriculture and Internal Improvement*. – London: G. & W. Nicholl, 1811. – xvi, 192p, plates: ill; 22cm.
Locations: EXU (RC 338.10942 WOR), PLU, WSL (630/COR/WOR)

Devon

Cooke, G. A. *A Topographical and Statistical Description of the County of Devon*. – 3rd edn – London: Sherwood, Neely & Jones, [1825?]. – *316p, plates: ill, map; 15cm.*
Locations: PLP, BRP, TQP, WSL(sDEV/1822/COO)

Cox, Thomas. *Magna Britannia; or, A Topographical, Historical, Ecclesiastical and Natural History of Devonshire*. – London: E. Nutt for M. Nutt, 1720. – pp. 465–546. – From his *Magna Britannia et Hibernia antiqua et nova; or, a new survey of Great Britain*.
Locations: PLP, WSL (sDEV/1720/COX)

Fraser, Robert. *General View of the County of Devon, with observations on the means of its improvement*, by Robert Fraser. – London: C. Macrae, 1794. – 70p; 4o.
Locations: WSL (s630/DEV/FRA), EXI, PLP

——*General View of the County of Devon, with observations on the means of its improvement*. – Barnstaple: Porcupines, 1970. – 75p, folded map; 25cm. – Facsimile of edn of 1793.
Locations: WSL (630/DEV/FRA), PLU, EXU (School of Ed 942.35/FRA)

Swete, John. Tour in North Devon c.1789.
Locations: DRO (Z19/2/17)

Tanner, H. *The farming of Devonshire*. – London, 1848.
Locations: EXI, PLP

Tanner, H. 'The farming of Devonshire', *Journal of the Royal Agricultural Society*, 9 (1849) 454–95.
Locations: EXU

Vancouver, Charles. *General View of the Agriculture of Devon, with observations on the means of its improvement, drawn up for the consideration of the Board of Agriculture.* – London: Richard Philips, 1808. – 479p, plates: ill; 23cm.
Locations: EXU, EXI, WSL (s630/DEV/VAN), PLP

——*General View of the Agriculture of Devon, with observation on the means of its improvement, drawn up for the consideration of the Board of Agriculture.* – Newton Abbot: David & Charles, 1969. – 479p, plates: ill; 23cm. – Facsimile reprint.
Locations: EXC, EXI, PLU, WSL (s630/DEV/VAN)

——*General View of the Agriculture of Devon, with observation on the means of its improvement, drawn up for the consideration of the Board of Agriculture.* – New York: A. M. Kelly, 1969. – 479p, plates: ill; 23cm. – Facsimile reprint.
Locations: WSL (s630/DEV/VAN)

White, William. *History, gazetteer and directory of the county of Devonshire, and the city and county of Exeter. With separate historical, statistical & topographical descriptions of all the boroughs, towns, ports, and bathing places.* – Sheffield: W. White, 1850. – 804p; 19cm. – There are large numbers of directories for each of the four South-West counties, but those of William White for Devon are among the best of the genre. A second edn was published for 1878–9 and a third for 1890.
Locations: WSL (380.1025/DEV/1850), BRP (380.1/B/1850), PLP (960.2), TQP, EXU (942.35 WHI), EXI (directories)

Williams, Thomas Hewitt. *Devonshire scenery; or, directions for visiting the most picturesque spots on the eastern and southern coast, from Sidmouth to Plymouth.* By a Devonian. – Exeter: Pollard, 1826.
Locations: EXI (AD02WIL)

——*A Guide to the Picturesque Scenery and Antiquities of Devonshire*, 2 vols. – Exeter, 1827–8.
Locations: WSL, PLP

——*Picturesque Excursions in Devonshire and Cornwall.* – London, 1804. Part I: Devonshire was the only part to be published.
Locations: WSL, PLP, TQP, WSL (sDEV/1804/WIL)

DORSET

Abbot, Andrew. Account of a five week holiday in Dorset. – 1798.
Locations: DRR (D/COL: F30)

Boswell, Edward. *The civil division of the county of Dorset, methodically digested . . . and a complete nomina villarum.* – Sherborne: W. Cruttwell, 1795. – [12], xxii, [10], 107p: map; 8o.
Locations: DOR, WSL (s352/DOR/BOS), YEO

——*The civil division of the county of Dorset, methodically digested . . . and a complete nomina villarum.* – 2nd edn – Dorchester: Weston [&c], 1833. – [6], viii, ii, ii, 192p: map; 23cm.
Locations: DOR, WSL (s352/DOR/BOS)

Claridge, John. *General View of the Agriculture in the County of Dorset, with observations on the means of its improvement, drawn up for the consideration of the Board of Agriculture.* – London: Printed by W. Smith, 1793. – [2], 49, [1]p; 4o.
Locations: DOR, EXU,WSL (s630/DOR/CLA)

Cox, Thomas. A *Compleat History of Dorsetshire.* – London: E. Nutt for M. Nutt, 1720. – pp. 548–605. From his *Magna Britannia et Hibernia antiqua et nova; or, a new survey of Great Britain.*
Locations: DOR (2 copies, both lacking title page), EXI (S48.15), SAS

Stevenson, W. *General View of the Agriculture in the County of Dorset, with Observations of the means of its Improvement.* – London, 1812. Reprinted in 1815.
Locations: EXU, DOR (1815 reprint), SAS (1812)

SOMERSET

Acland, Sir Thomas Dyke. *The farming of Somersetshire*, by Thomas Dyke Acland and W. Sturge. – London: John Murray, 1851. – xii, 179p: ill, maps; 23cm. Reprinted from: *Journal of the Royal Agricultural Society*, 11, part 2.
Locations: EXU, WSL (630/SOM/ACL), SAS, SSL

——*The Farming of Somersetshire*, 1850, / by T. D. Acland and W. Sturge. – Prize report reprinted from the *Journal of the Royal Agricultural Society*, 11, part 2.
Locations: EXU

Billingsley, John. *General View of the Agriculture in the County of Somerset . . . By John Billingsley. Drawn up for the consideration of the Board of Agriculture.* – London: W. Smith, 1794. – 192p; 4o.
Locations: SAS

——*General View of the Agriculture of the County of Somerset . . .*, now re-printed with considerable additions and amendments. – Bath: R. Cruttwell, 1797. – 320p; 8o.
Locations: EXI, SAS, SSL

——*General View of the Agriculture of the County of Somerset.* – 2nd edn – Bath: R. Cruttwell, 1798. – 320p; 8o.
Locations: EXU, PLU, SSL, YEO

Cooke, G.A. A *topographical and statistical description of the county of Somerset*. – London: Sherwood, Gilbert & Co., [1815]. – 208p, plates: ill, map; 15cm.
Locations: WSL (sSOM/1815/COO)

Cox, Thomas. *Magna Britannia; or, A Topographical, Historical, Ecclesiastical and Natural History of Somersetshire* – London: E. Nutt for M. Nutt, 1720. From his *Magna Britannia et Hibernia antiqua et nova; or, a new survey of Great Britain*.
Locations: EXI (S53.4) SAS, SSL.

——*A compleat history of Somersetshire . . .* – Sherborne, 1742. – 206p; 2o. – A reprint from Cox's *Magna Britannia*, issued in weekly parts with the *Sherborne Mercury* newspaper.
Locations: SAS, SSL

Greenwood, C. & J. *Somersetshire delineated: being a topographical description of each town, parish, chapelry, &c in the county . . . and useful as an index to the survey made in the year 1821*. – London: C. & J. Greenwood, 1822. – 215p; 21cm.
Locations: WSL (sSOM/1822/GRE), EXI (AS 03.1GRE), SAS, SSL, YEO

——*Somersetshire delineated: being a topographical description of each town, parish, chapelry, &c in the county*. – Gloucester: Alan Sutton, 1980. – [8], 215p; 23cm. – Facsimile reprint.
Locations: WSL (SOM/1822/GRE), SAS, SSL

Nightingale, Joseph. *Somersetshire; or, original delineations, topographical, historical, and descriptive, of that county. The result of a personal survey*. – London: Printed for J. Harris, 1818. – pp. 393–711, plate: ill. – Separate issue of part of *The beauties of England and Wales*.
Locations: WSL (sSOM/1818/NIG), SAS, SSL

Spurle, W. Diary of travels from Taunton to York, 1802 and to Bristol and Clifton, 1803. – 1802–3. – Written in the form of letters to his brother.
Locations: SRO (DD/X/HC/1)

WORKS COVERING MORE THAN ONE COUNTY

Anonymous. Journey from Cornwall to Birmingham. 1785. – 78p; 8o.
Locations: CRO (T 1341/1,2)

Anonymous, Journal of Tours in the Midland and Western Counties of England and Wales in 1794, and in Devonshire in 1803.
Locations: BL (Add. MS 30.172)

Anonymous. Journey from Cornwall to Exeter and Bristol. 1826.
Locations: CRO (X 20/40)

Ayton, Richard. A *Voyage round Great Britain undertaken in the summer of the year 1813 and commencing from Land's End, Cornwall.* – London, 1814–25. – 8 vols. – Illustrations by William Daniell.

——A *voyage round Great Britain undertaken between the years 1813 and 1823 and commencing from Land's End . . . with a series of views.* – London: Tate Gallery, 1978. – 2 vols, 149 leaves of plates; 24cm.
Locations: WSL (sWES/1813/AYT)

Banks, Sir Joseph. Journals of Excursions to Eastbury and Bristol, etc. 1767–8.
Locations: Cambridge University Library (Add. MS 6294)

Bridges, John. A *book of fairs: or, a new guide to West Country travellers: shewing them all the fairs in these thirteen several counties following viz. Gloucester-shire, Wilt-shire, Somerset-shire, Dorset-shire, Devon-shire, Cornwall, Glamorgan-shire, Monmouth-shire, Hereford-shire, Worcester-shire, Oxford-shire, Berk-shire, and Hamp-shire.* – The seventh impression. – London: Printed for the author, 1709.
Locations: BL (Cup 403.u.22)

Britton, John. T*he beauties of England and Wales*. – London: Vernor and Hood, 1801–15. – 18 vols – Vol 2: Cornwall, 1801, Vol. 4: Devon and Dorset, 1803, Vol. 13: Somerset, 1813. Individual sections were also reissued separately at later dates.
Locations: the reissues make this complex to list. Most major collections hold the sections for their individual counties, e.g CRO (VL/9/8, VL/9/12), WSL (sDEV/1803/BRI), SAS (Somerset)

Browne, E. Notes taken in a tour through the West of England, 1662.
Locations: BL (Sloane MS 1900)

Butcher, Edmund. *An excursion from Sidmouth to Chester, in the summer of*

1803, in a series of letters to a lady. – London: H. D. Symonds, 1805. – viii, 462p. – Devon, Somerset & Bristol pp. 1–82.
Locations: WSL (sWES/1803/BUT), SAS, SSL

Camden, William. *Britannia, sive florentissimorum regnorum Angliae, Scotiae, Hiberniae, et Insularum adjacentium ex intima antiquitate Chorograpica Descriptio.* – London: Bishop, 1594.
Locations: EXC (R.CAM), SAS

——*Britannia, sive florentissimorum regnorum Angliae, Scotiae, Hiberniae, et insularum adjacentium ex intima antiquitate chorograpica descriptio . . . nunc postremo recognita.* – Londini: impensis Georg. Bishop, 1600. – References: STC 4507.
Locations: EXU (Crediton Lib 1600/CAM)

——*Britannia: or, a chorographical description of the most flourishing kingdoms of England, Scotland and Ireland, and the islands adjacent, from the earliest antiquity,* translated from the ed published by the author in 1607. – London, 1789. – 3 vols.
Locations: PLU (Exmouth Campus)

——*Britannia: or, a chorographical description of the most flourishing kingdomes of England, Scotland and Ireland . . . beautified with mappes of the several shires of England,* translated newly into English by Philemon Holland . . . revised by the author. – London: F. K. R. Y and I. L. for George Latham, 1637. – 822, 233p. – Refs: STC 4510.
Locations: EXU (Rare Books B1637/CAM/X, Syon Abbey 1637/CAM/X)

——*Britannia: or, a Chorographicall description of Great Britain and Ireland,* translated into English with large additions by Edmund Gibson. – London, 1695.
Locations: EXC, SAS

——*Britannia: or, a Chorographicall description of Great Britain and Ireland,* translated into English by Edmund Gibson. – 2nd edn – London, 1722.
Locations: SAS

——*Britannia: or, a Chorographical description of Great Britain and Ireland,* 4th edn, translated into English by Edmund Gibson. – London, 1772.
Locations: EXU

——*Britannia: or, a Chorographical description of the most flourishing kingdoms of England, Scotland and Ireland, and the islands adjacent, from the earliest antiquity.* – London: J. Stockdale, 1806. – With a folding map by F. W. L. Stockdale, text enlarged by Richard Gough.
Locations: EXU, WSL (Devon section), SAS

——*Britannia*. – Newton Abbot, 1971. – Facsimile reprint of 1695 edn with an introduction by S. Piggott.
Locations: EXU

Clarke, Edward Daniel. *A tour through the south of England, Wales and part of Ireland, made during the summer of 1791*. – London: Minerva Press, 1793.
Locations: DOR (lacks 12 plates), EXI

Clutterbuck, R. Journal of a tour through the Western Counties of England during the summer of 1796.
Locations: Cardiff University Library (MS 3 277)

Cox, Thomas. *Magna Britannia antiqua & nova; or, a new, exact and comprehensive survey of the ancient and present state of Great-Britain*. – [London]: Printed by Eliz. Nutt and sold by M. Nutt and J. Morphew, 1720–31. – 6 vols, plates: ill, maps, 40. – First issued as a monthly supplement to *Atlas geographicus*, reissued for Caesar Ward 1738.
Locations: For county sections see under individual counties

Cullum, Sir John. Tour in the West of England, 1779.
Locations: Suffolk Record Office (E2/33/2)

Defoe, Daniel. *A Tour through England and Wales*. – London: Dent, 1927. – Everyman edition.
Locations: EXU, EXI

——*A Tour through England and Wales*. – London, 1928. – 2 vols. – Reprinted 1948.
Locations: EXU, PLU, PLP

——*A Tour through the whole island of Great Britain* . . . – 7th edn – London: J. & F. Rivington, 1769. – 4 vols. – 120.
Locations: SAS

——*A Tour through the whole island of Great Britain*, edited by G. D. H. Cole. – London: Dent, 1962.
Locations: DOR

——*A Tour through the whole island of Great Britain*, edited by Pat Rogers. – Harmondsworth: Penguin, 1971.
Locations: EXU, CRO (VL/9/2), SAS

——*A Tour through Great Britain*, edited by S. Richardson. – New York, 1975.
Locations: EXU

——*A Tour through the whole island of Great Britain*, abridged and edited by P. N. Furbank and W. R. Owens. – London: Yale University Press, 1991. – xiv, 417p: ill, maps; 28cm.
Locations: WSL (sxWES/1724/DEF)

Dunsford, Martin. *Miscellaneous observations in the course of two tours through several parts of the West of England*. – Tiverton: E. Boyce, 1800. – 152p; 17cm.
Locations: WSL, EXI

Early Tours in Devon and Cornwall/edited by R. Pearse Chope. – Newton Abbot: David & Charles, 1968. – First published as a supplement to *Devon and Cornwall Notes and Queries* in 1918. A series of extracts drawn from a selection of the works of well-known topographical authors. The versions are sometimes abridged and altered, but this nonetheless remains a relatively useful, accessible volume.
Locations: WSL

Enys, John. Journey from Yorkshire to Cornwall. – 1783/4. – 26p; 4o.
Locations: CRO (EN1806)

Ernst, Elizabeth. Journal of tours in Somerset and Devon, 26 July–11 Aug. 1843, 8–15 July and 6–22 Sept 1845. – 1843–45.
Locations: SRO (DD/SWD/10/9 & 11)

Ernst, T.H. Journal of a tour in Surrey, Hants, Somerset and Devon, 28 Sept –29 Oct 1813. – 1813.
Locations: SRO (DD/SWD/10/1)

Fiennes, Celia. *The Journeys of Celia Fiennes*, edited with an introduction by C. Morris. – London: Cresset Press, 1947. – xiv, 376p: maps; 19cm.
Locations: EXU, EXC, WSL (WES/1685/FIE)

——*The Journeys of Celia Fiennes*, edited with an introduction by C. Morris. – New edn – London, 1949.
Locations: EXU, EXI, PLP, CRO (VL/9/1), DOR, SSL

——*The Illustrated Journeys of Celia Fiennes*, c.1682–c.1712. – 2nd ed., edited by C. Morris. – London: Macdonald, 1982. – 248p: ill.
Locations: WSL (sWES/1682/FIE), EXI, PLU

Fuller, T. *The History of the Worthies of England*, published in 1662. – New edn, edited by J. Nichols. – Rivington, 1811.
Locations: EXI, SAS

——*The History of the Worthies of England*, published in 1662. – New edn – London, 1840. – 3 vols.
Locations: EXC

——*The Worthies of England*, edited with an introduction and notes by J. Freeman. – London, 1952.
Locations: EXU, PLU (Exmouth Campus)

Gilpin, William. *Observations on the Western Parts of England relating chiefly to the picturesque beauty*. – London: T. Cadell & W. Davies, 1798.
Locations: BRP, TQP, WSL (sWES/1798/GIL), PLP, CRO (VL/9/4), DOR (1808 edn)

——Observations on the Western Parts of England relating chiefly to the picturesque beauty, by William Gilpin; new introduction by Sutherland Lyall. – Richmond: Richmond Publishing, 1973. – 359p, plates: ill; 23cm.
Locations: WSL (WES/1798/GIL), SSL

Grey, Sir Thomas. Tour in Devonshire and the West Country, 1789.
Locations: Suffolk Record Office (E2/44/6)

Grose, Francis. *The antiquities of England and Wales*. – London: Printed for S. Hooper, 1772–1787. – 6 vols, plates: maps, ill.
Locations: EXI (Devon section: S31.1–8), WSL (Devon section pxDEV/1785/GRO)

Hammond, Lieutenant. A relation of a short survey of the Western Counties made by a Lieutenant Hammond of the Military Company in Norwich. – 1635.
Locations: BL (Lansdowne MS 213/27)

——*A relation of a short survey of the Western Counties made by a Lieutenant Hammond of the Military Company in Norwich*, in 1635, edited by L. G. Wickham Legg. London: Camden Society, 1936. – (Camden Miscellany 16; Camden Society, 3rd Series, 52). – xiv, 128p.
Locations: WSL (WES/1635/CAM)

Harrison, W. *Description of England*, written in the 1560s or 1570s, re-published as *Description of England by William Harrison*, edited by G. Edelen. – Cornell, 1968.
Locations: EXU

Jones, William. *A complete history of all the religious houses in the counties of Devon and Cornwall before the Dissolution*. – London: Smerdon and Underhill, 1779. – viii, 88p; 8o.
Locations: WSL (s271/WES/JON)

Leland, John. *Johannis Lelandi antiquarii de rebus Britannicis Collectanea cum Thomae Hearnii*, 2nd edn with additions. – London: Richardson, 1770.
Locations: EXI

——*Johannis Lelandi antiquarii de rebus Britannicis Collectanea*, edited by T. Hearne. – London, 1774. – 6 vols.
Locations: PLP

——*The Itinerary of John Leland the Antiquary*. – 2nd edn – Oxford, 1745. – 9 vols.
Locations: SAS

——*The Itinerary of John Leland the Antiquary*, to which is prefixed Mr Leland's New Year gift. – 3rd edn – Oxford: J. Fletcher, 1768.
Locations: EXI (S 70)

——*The Itinerary of John Leland the Antiquary*. – Oxford, 1769.
Locations: EXC

——*The Itinerary of John Leland in or about the years 1535–1543*, 5 vols, edited by L. Toulmin Smith. – London: Bell, 1906.
Locations: EXU, EXI (H 05 LEL), PLU, PLP, SAS (1964 reprint)

Lipscomb, George. *A Journey into Cornwall from Hampshire through the Counties of Southampton, Wiltshire, Dorset, Somerset and Devon, Warwick, interspersed with remarks moral, historical, literary and political*. – Warwick: H. Sharpe, 1799.
Locations: EXI, PLP, WSL (sWES/1799/LIP), CRO (VL/9/5), DOR

Magalotti, L. *Travels of Cosmo the Third, Grand Duke of Tuscany, through England during the reign of Charles II, 1669*, translated from the Italian Manuscript in the Laurentian Library, Florence. – London, 1821. – Devon and Cornwall sections reprinted in *Early tours in Devon and Cornwall*.

Marshall, William. *The Rural Economy of the West of England*. – London: G. Nicol, 1796. – 2 vols.
Locations: EXU, EXC, EXI, PLP, WSL (s630.2/WES/MAR)

——*The Rural Economy of the West of England*. – 2nd ed. – London: G. & W. Nicol, 1796. – 2 vols: map; 23cm.
Locations: WSL (s630/WES/MAR)

——*Marshall's Rural Economy of the West of England* (1796). – Newton Abbot: David & Charles, 1970. – 2 vols: folded map; 22cm. – Facsimile reprint.
Locations: PLU, PLP, WSL (630/WES/MAR), DOR, SSL

Martin, Benjamin. The natural history of England. – London: W. Owen, 1759–63. – 2 vols, plates: ill, maps ; 8°. – Previously issued as part of Martin's *The general magazine of arts and sciences*.
Locations: WSL (s500.9/WES/MAR), Bristol Univ (Somerset only)

Maton, W.G. *Observations relative chiefly to the natural history, picturesque scenery and antiquities of the Western Counties of England, made in the years 1794 and 1796*. – Salisbury, 1797. – 2 vols.
Locations: EXI, PLP, TQP, CRO (VL/9/3), SAS

Milles, Jeremiah. Summer Tours to Wales and the Southern Counties, 1735–43.
Locations: BL (Add. MS 15,776)

Murray, John. A *handbook for Travellers in Devon and Cornwall*. – London: John Murray, 1851. – 243p: folded map; 18cm.
Locations: WSL (WES/1851/HAN), CRO (VL/9/18)
Later editions up to the 11th (1985) are to be found in many libraries, both as combined volumes and as separate volumes for the individual counties.

——A *Handbook for Travellers in Devon and Cornwall*. – Newton Abbot: David & Charles, 1971. – 282p: folded map; 19cm. – Facsimile reprint of 4th edn, published in 1859.
Locations: EXU, EXI, CAM, WSL (WES/1859/MUR)

——A *Handbook for Travellers in Wiltshire, Dorsetshire and Somersetshire*. – 4th edn – London, 1882.
Locations: EXI
Other editions can be found in various libraries in the region.

Osborne, Francis Godolphin, fifth Duke of Leeds. Tour to the West in 1791, accompanied by the Duchess and others.
Locations: BL (Add. MS 28,570)

Pococke, R. *The Travels through England of Dr Richard Pococke, successively Bishop of Meath and of Ossory during 1750, 1751, and later years*, edited by J. J. Cartwright. – London: Camden Society, 1888–89. – 2 vols. – (Camden Society. Second series).
Locations: EXU, EXI

Punchard, F. 'Farming in Devon and Cornwall', *Journal of the Royal Agricultural Society*, 51 (1890), 511–36
Locations: EXU

Rackett, Rev. Thomas. An Account of a Journey through Devon and Cornwall (nineteenth century).
Locations: DRR (D/RAC: NU 181)

Rackett, Mrs Thomas. An Account of a Journey through Devon and Cornwall (nineteenth century).
Locations: DRR (D/RAC: NU 140)

Roberts, George. *The social history of the people of the southern counties of England in past centuries*. – London: Longman, Brown, Green, Longmans and Roberts, 1856. – xvi, 572p; 22cm.
Locations: DOR

Rural elegance display'd, in the description of four western counties: Cornwall, Devonshire, Dorsetshire, and Somersetshire. – London: Printed for Staples Stearne, 1768. – xxiv, 314p; 12o.
Locations: sWES/1768/RUR)

Shaw, S. A *Tour in the West of England in 1788*, / by the Rev. S. Shaw, M.A., Fellow of Queen's College, Cambridge – London: for Robson & Clarke, 1789. – viii, 602p; 8o.
Locations: WSL (sWES/1788/SHA), EXI, DOR, SAS, SSL

Skinner, John. Copy of a Journal kept by the Revd John Skinner of a Tour to Wells, Bridgwater, Taunton, Exeter, Sidmouth, Okehampton, and in Cornwall from 20 Sept to 25 Nov. 1797.
Locations: BL (Add. MS 28,793), WSL (microfilm)

——Letters on antiquities. – 1815–29.
Letters to and from Lysons, Rev James Douglas and Sir Richard Colt Hoare.
Locations: BL (Add. MS 33665), WSL (microfilm)

——Journal of a Western tour. – 1829.
Locations: BL (Add. MS 33714), WSL (microfilm)

——Diary of a tour through part of Somersetshire, Devonshire and Cornwall in the autumn of 1797. – 1797.
Locations: SRO (DD/SAS/C/1193/10a)

——A tour through parts of Somersetshire and Devonshire to Sidmouth; and from thence along the southern coast, through Dorsetshire, Hampshire and Sussex, to Broadstairs in Kent. – 1801.
Locations: SRO (DD/SAS/C/1193/10d)

——*West Country tour: being the diary of a tour through the counties of Somerset, Devon and Cornwall in 1797*. – Bradford-on-Avon: Ex Libris, 1985. – 96p: ill, map; 18cm.
Locations: WSL (WES/1797/SKI), SAS, SSL

Spoure, Edmund. Journey to Bath and Oxford. – 1693.
Locations: CRO (FS3/93/8/297–312)

Stackhouse, Rachel. Visit to Bath and Salisbury. – 1800.
Locations: CRO (RS/1/174)

Stukeley, William. *Itinerarum Curiosum*. – London, 1724.

——*Itinerarum Curiosum*, 2nd edn, with large additions. – London, 1776.
Locations: EXU, SSL (1969 reprint)

Revd John Swete, Travel Journals, 1789–1804. – 17 vols.
Locations: DRO (564M/F1–F17)

Tunnicliff, W.A. *Topographical Survey of the Counties of Somerset, Gloucester, Worcester, Stafford, Chester and Lancaster*. – Bath, 1789.
Locations: EXI, SSL

——*Topographical Survey of Hants, Wilts, Dorset, Somerset, Devon and Cornwall*. – Salisbury: Collins, 1791.
Locations: EXI, DOR

Walcott, Mackenzie. A guide to the coasts of Devon and Cornwall descriptive of scenery historical, legendary and archaeological. – London: Edward Stanford, 1859.
Locations: WSL (sWES/1854/WAL), BRP (A/1854/WAL)

Warner, Richard. A *walk through some of the western counties of England*. – Bath: Printed by R. Cruttwell, 1800. – vi, 222p, 2 leaves of plates: ill, maps, 8o.
Locations: WSL (sWES/1800/WAR), SAS, SSL

White, Walter. A Londoner's walk to the Land's End; and a trip to the Scilly Isles. – London: Chapman & Hall, 1855. – *x, 357p: folded map; 20cm*.
Locations: WSL (WES/1855/WHI), EXU (RC 942.37)

William of Worcester. Itineraria Symonis Simeonis et Willelmi de Worcestre. 1478.
Locations: Corpus Christi College Library, Cambridge (MS 210)

——*William Worcestre: Itineraries*, edited by J.H. Harvey. – Oxford, 1969.
Locations: SSL

——*Itineraria Symonis Simeonis et Willelmi de Worcestre*, edited by J. Nasmith. – Cambridge, 1778.
Locations: EXI

Windham, W. Journal of a Tour through Surrey, Wiltshire, Dorset, Devon, Somerset, Wales, Worcestershire, etc. 7 June–21 August 1779.
Locations: BL (Add. MS 37,926)

Wynne, Luttrell. Sketchbooks. – 1771. – Covers Cornwall, Devon, Oxfordshire, Derbyshire, Warwickshire.
Locations: CRO (PD439)

Wynne, William. Journey from London to Cornwall. – 1755. – 30p; 40.
Refs: Cornish section edited by J. C. Edwards in *Journal of the Royal Institution of Cornwall*, new series, 8, pt. 4 (1981).
Locations: CRO (PD 220)

Section 3: Bibliographies and Guides to Research

Catalogues of printed books and other material relating to the history and topography of each of the four counties of South-West England were produced in the nineteenth century or early twentieth century. These remain useful works of reference for today's researcher.

Cornwall

Boase, George Clement and Courtney, William Prideaux. *Bibliotheca Cornubiensis: A catalogue of the writings, both manuscript and printed of Cornishmen and the works relating to the county of Cornwall with biographical memoranda and copious literary references*. – London: Longmans, Green, Reader & Dyer, 1874. – 3 vols. – Vol. 1–2: alphabetical list of items. Vol. 3: supplementary catalogue listing Acts of Parliament and Civil War tracts with an index to all three volumes.
Locations: WSL (sx016/COR/BOA)

Devon

Davidson, James. *Bibliotheca Devoniensis: A catalogue of the printed books relating to the county of Devon*. – Exeter: W. Roberts, 1852. – [5], 226p. – Intended

for 'those who take an interest in the Topography, History or Biography of the County'. Thematic sections on general and local history, political history, ecclesiastical history, agriculture, the stannaries, manufacturing, medicine, natural history, poetry, biography and family history. List of periodical works and local acts of Parliament. Indexes to places and another to persons.
Locations: WSL (016/DEV/DAV, s016/DEV/DAV Inv: S3232 Davidson's copy, extensively annotated)

——*Supplement to the Bibliotheca Devoniensis*. – Exeter: [s.n.], 1861. – 51p.
Locations: WSL (s016/DEV/DAV Davidson's copy, extensively annotated)

Somers Cocks, J.V. *Devon Topographical Prints, 1660–1870: A catalogue and guide*. – Exeter: Devon Library Services, 1977. – viii, 324p. – Brief but useful introduction which discusses the history of print making and print selling, artists and engravers, the development of 'vignette views' between 1841 and 1876, and publishers and publishing. Catalogue arranged alphabetically and since many of the prints were published as illustrations in a topographical text, it provides a very considerable amount of detail and information in addition to that purely about the artwork.
Locations: WSL (016.7694991/DEV/SOM), BRP, PLP, TQP, SSL

Brockett, Allan. *The Devon Union List (D.U.L.): A collection of written material relating to the county of Devon*. – Exeter: University of Exeter Library, 1977. – iii, 571p. – Lists books, pamphlets and a small selection of other material on Devon which is held in six major libraries in the county. The level of information is somewhat skeletal, but in providing the location of the items, Brockett's work is extremely useful.
Locations: WSL (016/DEV/BRO), BRP, PLP, TQP, SSL

Dorset

Mayo, Charles Herbert. *Bibliotheca Dorsetiensis: Being a carefully compiled account of printed books and pamphlets relating to the history and topography of the county of Dorset*. – London: Chiswick Press, 1885. – 296p. – Thematic sections including antiquarian literature, historical literature, political pamphlets, ecclesiastical literature, works illustrating social life, the Dorset dialect, agricultural publications, natural histories, newspapers, Acts of Parliament, maps of Dorset, and works relating to particular parishes. Section on the writings of those who travelled in the county, and another on Dorset guide books.
Locations: DOR, WSL (s016/DOR/MAY)

Douch, Robert. A *Handbook of Local History: Dorset*. – 2nd edn – Bristol: University of Bristol, 1962. – An invaluable source on the topographical and historical works on Dorset, first published in 1952. A listing of of both general works and other sources of information is provided together with details of repositories holding important collections of material. The final chapter deals with works on particular places and the book contains an index to Dorset places.
Locations: DOR, WSL (s016/DOR/DOU, 1952 ed), DRR (1952 edn), SSL

Carter, Kenneth. *Dorset: a catalogue of the books and other printed material . . . in Dorset County Library*. – Dorchester: Dorset County Council, 1974. – viii, 183p.
Locations: WSL (s016/DOR/DOR), DOR, SSL

SOMERSET

Green, Emanuel. *Bibliotheca Somersetensis: A Catalogue of Books, Pamphlets, single sheets, and Broadsides in some way connected with the County of Somerset*. – Taunton: Barnicott & Pearce, 1902. – 3 vols. – Vol. 1: general introduction and a listing of books on Bath. Vols 2–3: "County" books in alphabetical order. His thorough approach frequently provides a reconstruction of the publication history of a particular work. There is an extensive and valuable listing of Civil War tracts relating to Somerset and also a listing of the papers and publications of the Bath and West of England Society (for the Encouragement of Agriculture, Arts, Manufactures and Commerce).
Locations: WSL (s016/SOM/GRE)SAS, SSL, YEO

SECTION 4: GENERAL CATALOGUES AND LISTINGS

A range of other catalogues and listings offer further guidance on the topographical sources available for South-Western counties. The most accessible of these include the following:

Anderson, J.P. *The Book of British Topography: A classified catalogue of the topographical works in the Library of the British Museum relating to Great Britain and Ireland*. – London, 1881.

Aubin, R.A. *Topographical poetry in XVIIIth century England*. – New York, 1936.

Batts, John Stuart. *British manuscript diaries of the 19th century: an annotated listing*. – Centaur Press, 1976. – Chronological listing which includes many minor and anonymous figures with topographical index.

Bayne-Powell, R. *Travellers in eighteenth century England.* – London, 1951.

Colt Hoare, R. A *catalogue of books relating to the history and topography of England, Wales, Scotland, and Ireland . . . compiled from the Library at Stourhead.* – London, 1815.

English County Histories: a Guide, A Tribute to C. R. Elrington, edited by C. R. J. Currie and C. P. Lewis. – Stroud, 1994. – Produced as a festschrift to Christopher Elrington, former editor of the Victoria County History. A brief essay for every English county, each by a different author, introducing the key historians, though the approach varies considerably from one contributor to another. There is a short chapter on Cornwall by John Walker, another on Devon by Joyce Youings, Dorset is covered by J. H. Bettey, and Somerset by Robert Dunning, but these represent only a fraction of a very large work and their contents overlap very little with the chapters presented in this book.

Gard, R. (ed) *The Observant Traveller: Diaries of Travel in England, Wales and Scotland in the County Record Offices of England and Wales,* Association of County Archivists. – London, 1989. – Illustrated volume containing a series of extracts covering subjects discussed by diarists, such as the journey, lodgings and refeshment, and taking the Waters. It is by no means comprehensive and the number of entries referring to locations in the South West is disappointing. However, a useful catalogue of (mostly eighteenth and nineteenth century) diaries, arranged by county, is included at the end.

Gough, R. A *Catalogue of books relating to British Topography and Saxon and Northern Literature, bequeathed to the Bodleian Library in 1799.* – Oxford, 1814.

Gough, R. *British Topography: or an Historical Account of what has been done illustrating the Topographical Antiquities of Great Britain and Ireland.* – London, 1780.

Malcolm, J. P. *Lives of Topographers and Antiquaries who have written concerning the Antiquities of England with (twenty-six) Portraits of the Authors, and a complete list of their works so far as they relate to the Topography of this Kingdom.* – London, 1815.

Matthews, W. *British Diaries: An Annotated Bibilography of British Diaries Written between 1442 and 1942.* – Berkeley, California, 1950. – Though very useful, this bibliography is arranged chronologically rather than geographically, and the index is merely an alphabetical list of the diarists. There is no rapid way to identify diaries which refer particularly to South-West England. However, Matthews provides a useful starting point for a researcher interested in this kind of material and his listing is much less idiosyncratic in its coverage than that edited by Robin Gard.

Moir, E. *The Discovery of Britain: The English Tourists*. – London, 1964. – Moir's book includes a lengthy bibliography (pp. 157–78) where she identifies a large corpus of printed and manuscript travel literature which contains a great deal of topographical description. Thus, many items which are cited are relevant to the study of past topographical works on the South-Western counties.

Nichols, J. *Bibliotheca Topographica Britannica*. – London, 1780–90. – 10 vols.

Piggott, S. *Ruins in a Landscape: Essays in Antiquarianism*. – Edinburgh, 1976. – A compendium of Piggott's essays published during the previous 25 years. There is therefore considerable overlap and repetition. But this is a thought-provoking work which also contains bibliographical material.

Upcott, W. *A bibliographical account of the principal works relating to English topography*. – London: R. & A.Taylor, 1818. – 3 vols.

——*A bibliographical account of the principal works relating to English topography*, with a new introduction by J. Simmons: Wakefield: E.P. 1978. – Lists catalogues of topographical works, general topographical studies, and topographical works on English counties, arranged alphabetically. Its value is enhanced by the detailed information on the content of each book in the listing, including the contents of J. Nichols' *Bibliotheca Topographica Britannica*. Index of places.
Locations: WSL (s016/WES/UPC)

Worrall, J. *Bibliotheca Topographica Anglicana: A Catalogue of Books on English Topography*. – London, 1736.

Chapter Notes

1 Introduction: The Development of Topographical Writing in the South West

1. R. Carew, *The Survey of Cornwall*, new edn (London, 1769) preface 'to the reader' following xxv; see also R. Carew, *The Survey of Cornwall, written by Richard Carew of Antonie, Esq.* (London, printed by S. S. for John Jaggard 1602).
2. E. Moir, *The Discovery of Britain: The English Tourists* (London, 1964), 12–23; W. G. Hoskins, *Provincial England: Essays in Social and Economic History* (London, 1965), 209–29; J. Youings, *Sixteenth-Century England*, Penguin Social History of Britain (Harmondsworth, 1984), 356–8; S. Piggott, *Ruins in a Landscape: Essays in Antiquarianism* (Edinburgh, 1976), 33–4.
3. Hoskins, *Provincial England*, 209. Chorography was taken by Tudor and Stuart writers to mean an investigation of the areal differentiation of the earth. Bernhard Varenius (author of *Geographia generalis*, published in 1622) distinguished between topographical investigations, which focused on a small tract or place, and chorographical studies which dealt with medium-sized regions. See J. N. L. Baker, 'The geography of Bernhard Varenius', *Transactions of the Institute of British Geographers*, 21 (1955), 51–60.
4. Hoskins, Ibid. 209–11; M. W. Greenslade, 'Introduction: county history', in C. R. J. Currie and C. P. Lewis (eds) *English County Histories: A Guide* (Stroud, 1994), 14–16.
5. Greenslade, 'Introduction: county history', 20.
6. British Library, Additional MS 31853, f. 2v; British Library, Harleian MS 6252; J. Norden, *Speculi Britanniae Pars: A Topographical and Historical Description of Cornwall* (1728, Frank Graham reprint 1966). See also W. Ravenhill, 'The missing maps from John Norden's Survey of Cornwall', in K. J. Gregory and W. L. D. Ravenhill (eds) *Exeter Essays in Geography* (Exeter, 1971), 93–104.
7. W. Lambarde, *Perambulation of Kent* (London, 1576); quoted by J. Youings, *Sixteenth-Century England*, 358.
8. Greenslade, 'Introduction: county history', 15.
9. Hoskins, *Provincial England*, 215.

10. Ibid. 216; see also S. Reynolds, 'Christopher Elrington and the V.C.H.', in Currie and Lewis, *English County Histories*, 1–4; Greenslade, 'Introduction: county history', 22–5.
11. R. N. Worth, 'William of Worcester: Devon's earliest topographer', *Transactions of the Devonshire Association*, 18 (1886), 485. Robert Bracey was his cousin.
12. Ibid. 487.
13. J. Nasmith, *Itineraria Symonis Simeonis et Willelmi de Worcestre* (London, 1778); Corpus Christi College Library, Cambridge, MS 210. See also J. H. Harvey (ed.) *William Worcestre itineraries* (Oxford Medieval Texts, 1969).
14. Moir, *Discovery of Britain*, 12; see also E. Burton, *The Life of John Leland, the first English antiquary, with a bibliography of his works* (London, 1896); J. Chandler, *John Leland's Itinerary: Travels in Tudor England* (Stroud, 1993), an indispensible new guide to Leland's works.
15. Quoted by W. Ravenhill, *John Norden's Manuscript Maps of Cornwall* (Exeter, 1972), 11; by Hoskins, *Provincial England*, 210; and by Moir, *Discovery of Britain*, 16–17.
16. G. Edelen (ed.) *Description of England by William Harrison* (Cornell, 1968).
17. Ravenhill, *John Norden's Manuscript Maps of Cornwall*, 11.
18. W. Camden, *Britannia, sive Forentissimorum Regnorum Angliae, Scotiae, Hiberniae, et Insularum adjacentium ex intima antiquitate Chorographica Descriptio* (London, 1586); see also S. Piggott, 'William Camden and the *Britannia*', in Piggott, *Ruins in a Landscape*, 33–54. For a brief comment on Camden's contribution to Devon archaeology see J. Bosanko, 'Antiquarians and archaeologists: the history of archaeology in Devon', in *Archaeology of the Devon Landscape: Essays on Devon's Archaeological Heritage*, Devon County Council (Exeter, 1980), 13–21.
19. Greenslade, 'Introduction: county history', 12–14; see also Youings, *Sixteenth-Century England*, 358.
20. Greenslade, 'Introduction: county history', 14.
21. S. Piggott, 'Introduction' to facsimile edition of Gibson's 1695 edition of *Britannia* (Newton Abbot, 1971), 9; see also R. A. Butlin, 'Regions in England and Wales c.1600–1914', in R. A. Dodgshon and R. A. Butlin (eds) *An Historical Geography of England and Wales*, 2nd edn (London, 1990), 229.
22. Piggott, 'Introduction' to facsimile edition of Gibson's 1695 edition of *Britannia*.
23. Camden, *Britannia* (London, 1722 edn), 'Mr Camden's preface', i–ii.
24. Youings, *Sixteenth-Century England*, 358. For an account of the activities of the Elizabethan College or Society of Antiquaries see also R. Carew, *The Survey of Cornwall, and an Epistle concerning the Excellencies of the English Tongue. With a Life of the Author by H**** C***** Esq*, new edn (London, 1769), xviii–xix.
25. F. V. Emery, 'English regional studies from Aubrey to Defoe', *Geographical Journal*, 124 (1958), 315; see also F. E. Halliday (ed.) *Richard Carew of Antony: The Survey of Cornwall* (London, 1953), 11.
26. BL, Harleian MS 5827; Devon Record Office [DRO], H 783; see also W. J. Blake, 'Hooker's Synopsis Corographical of Devonshire', *Transactions of the Devonshire Association*, 47 (1915), 334–68.

27. E. H. Bates (ed.) *The Particular Description of the County of Somerset*, Somerset Record Society, xv (1915); R. Legg (ed.) *Thomas Gerard's General Description of Dorset* (Sherborne, Dorset 1980). As Joseph Bettey notes in his chapter, Thomas Gerard's manuscripts were published in 1732 and mistakenly attributed to John Coker, the rector of Mappowder: J. Coker, *A Survey of Dorsetshire, Publish'd from an Original Manuscript, written by the Reverend Mr Coker of Mapowder* (London, 1732).
28. T. Fuller, *The Worthies of England*, ed. by J. Freeman (London, 1952).
29. The Westcountry Studies Library in Exeter holds several manuscript versions of Risdon's 'Chorographical Description' and there is another in the West Devon Record Office. Further details of these are provided in the Listing at the end of the book. See also T. Risdon, *Chorographical Description or Survey of the county of Devon*, ed. by E. Curll (London, 1714). There are several later versions of Tristram Risdon's study. T. Westcote, *A View of Devonshire in MDCXXX, with a pedigree of most of its gentry*, ed. by Revd George Oliver and Pitman Jones (Exeter, 1845).
30. W. Pole, *Collections towards a description of the county of Devon* (London, 1791).
31. Emery, 'English regional studies', 315–17; Piggott, *Ruins in a Landscape*, 101–15; Butlin, 'Regions in England and Wales', 229–30.
32. Piggott, *Ruins in a Landscape*, 101–2; for the ideas and activities of the Royal Society in its early years see T. Birch, *History of the Royal Society of London*, 4 vols (London, 1756).
33. Piggott, ibid. 102; Emery, 'English regional studies', 315.
34. J. Aubrey, *The Natural History of Wiltshire*, ed. by J. Britton, Wiltshire Topographical Society (London, 1847); this misses out substantial sections of Aubrey's manuscript, in particular the classification of Wiltshire soils. For a fuller discussion of John Aubrey's work see K. Rogers and D. Crowley, 'Wiltshire', in Currie and Lewis (eds) *English County Histories*, 411–13.
35. Bodleian, MS Aubrey 4, f. 243; see also E. G. R. Taylor, 'Robert Hooke and the cartographical projects of the late seventeenth century', *Geographical Journal*, 90 (1937), 532–4.
36. R. Plot, *Natural History of Staffordshire* (Oxford, 1686); R. Plot, *Natural History of Oxfordshire* (London, 1705).
37. Plot, *Staffordshire*, 392. The term 'chorography' was also accorded a new, more restricted definition by the seventeenth-century natural historians who used it to mean variations in soils, geology and vegetation.
38. Emery, 'English regional studies', 320–1.
39. BL, Add. MS 27763; see also J. Walker, 'Cornwall', in Currie and Lewis (eds), *English County Histories*, 87. Much of Tonkin's work was reprinted in nineteenth-century parochial histories.
40. Walker, 'Cornwall', 87; H. L. Douch, 'Thomas Tonkin: an appreciation of a neglected historian', *Journal of the Royal Institution of Cornwall*, new series, 4 (1962), 145–80; F. A. Turk, 'Natural History Studies in Cornwall, 1700–1900', *Journal of the Royal Institution of Cornwall*, new series, 3 (1959), 229–79.

41. BL, Add. MS 29,762; W. Hals, *Compleat History of Cornwall* (Truro, 1750).
42. W. Borlase, *Natural History of Cornwall* (Oxford, 1758).
43. W. Borlase, *Observations on Antiquities Historical and Monumental of Cornwall* (Oxford, 1754).
44. Quoted in Borlase, *Antiquities Historical and Monumental*, facsimile reprint of 1763 edn by E. P. Publishing, with an introduction by P. A. S. Pool and C. Thomas (Wakefield, 1973). See also the list of William Borlase manuscripts listed in the Appendix to P. A. S. Pool, *William Borlase* (Truro, 1986), 285–97.
45. Somerset Record Office, DD/SH 96–8; see also R. W. Dunning, 'Somerset', in Currie and Lewis (eds), *English County Histories*, 348–9.
46. J. Collinson, *History and Antiquities of Somerset, collected from authentick records and an actual survey made by the late* Mr. *Edmund Rack. Adorned with a map of the county, engravings of Roman and other reliques, town seals, baths, churches and gentlemen's seats*, 3 vols (Bath, 1791). Collinson's proposals for printing 'by subscription the history of the county of Somerset' were published in Bath in 1784. A more accessible version is J. Collinson, *History & Antiquities of the County of Somerset*, with a new introduction by Robert Dunning (Gloucester, 1983).
47. T. Cox, A *Compleat History of Dorsetshire, being pages 548–605 inclusive of Magna Britannia et Hibernia Antiqua st Nova, or a New Survey of Great Britain in 6 volumes* (London, 1720–31).
48. J. Hutchins, *The History and Antiquities of the County of Dorset*, 2 vols (London, 1774).
49. J. Youings, 'Devon', in Currie and Lewis (eds), *English County Histories*, 120.
50. R. Polwhele, *History of Devonshire*, 3 vols (Exeter, 1793–1806). A facsimile reprint of this work was published for the Scolar Press by Kohler and Coombes (Dorking, 1978) and this includes an introduction by A. L. Rowse and plates of the handwritten index prepared by James Davidson.
51. R. Polwhele, *History of Cornwall*, 3 vols (Falmouth, 1803); enlarged edn in 7 vols (London, 1816). A facsimile reproduction of the 7 volume work was published in 3 volumes for the Scolar Press by Kohler and Coombes (Dorking, 1978) and this includes a brief introduction by A. L. Rowse.
52. D. Lysons and S. Lysons, *Magna Britannia, being a concise topographical account of the several counties of Great Britain, volume the third containing Cornwall* (London, 1814).
53. D. Lysons and S. Lysons, *Magna Britannia, being a concise topographical account of the several counties of Great Britain, volume the sixth containing Devonshire* (London, 1822): Daniel Lysons' preface.
54. C. S. Gilbert, *Historical Survey of the County of Cornwall*, 2 vols (Truro, 1817–20).
55. Walker, 'Cornwall', 88.
56. T. Moore, *The history and topography of the county of Devon including outlines of the physical geography, geology and natural history of the county*, 2 vols (London, 1831).
57. D. Gilbert, *The Parochial History of Cornwall, founded on the Manuscript Histories of Mr. Hals and Mr. Tonkin*, 4 vols (London, 1838).
58. J. Polsue, *Complete Parochial History of Cornwall*, 4 vols (Truro, 1867–72).

59. S. Rowe, *A Perambulation of the Ancient Royal Forest of Dartmoor and the Venville Parishes: Or, A Topographical Survey of their Antiquities and Scenery* (Exeter, 1848). This work was reprinted in 1985 as a facsimile by Devon Books, Exeter.
60. I. Stoyle, 'An overlooked description of Crediton', *Devon Historian*, 35 (1985), 18–22.
61. T. Gray, *The Garden History of Devon: An Illustrated Guide to the Sources* (Exeter, 1995). This contains detailed references to numerous diaries and travel journals. See also W. Matthews, *British Diaries, 1442–1942* (Berkeley, California, 1950); R. Pearse Chope, *Early Tours in Devon and Cornwall* (New York, 1968) reprint of 1918 edn with an introduction by A. Gibson; R. Gard, *The Observant Traveller: Diaries of Travel in England, Wales and Scotland in the County Record Offices of England and Wales* (London, 1989); Moir, *Discovery of Britain*, 158–78; G. E. Fussell and C. Goodman, 'Travel and topography in eighteenth century England', *Transactions of the Bibliographical Society*, 2nd series, 10 (1930), 84–103; G. E. Fussell and C. Goodman, 'Travel and topography in seventeenth century England', Ibid. 13 (1930), 292–311.
62. L. G. Wickham Legg [Hammond], *A Relation of A short Survey of the Western Counties made by a Lieutenant of the Military Company in Norwich in 1635*, Camden Miscellany, XVI (1936).
63. [Count Lorenzo Magalotti], *Travels of Cosmo the Third, Grand Duke of Tuscany, through England during the reign of Charles II, 1669, Translated from the Italian Manuscript in the Laurentian Library, Florence* (London, 1821). The Cornwall and Devon section is also available in Pearse Chope, *Early Tours*, 92–111.
64. C. Morris (ed.), *The Illustrated Journeys of Celia Fiennes c.1682 – c.1712* (Exeter, 1982), 36–47. For a discussion of Celia Fiennes and Daniel Defoe see Moir, *Discovery of Britain*, 35–46.
65. Morris, *Celia Fiennes*, 41.
66. Ibid. 200.
67. Ibid. 32.
68. Moir, *Discovery of Britain*, 35. See also D. Defoe, *A Tour Through the Whole Island of Great Britain*, Penguin English Library edn (Harmondsworth, 1971), 11–16. There are edited extracts taken from the Cornwall and Devon passages of both Fiennes and Defoe in Pearse Chope, *Early Tours*, 111–37 and 145–78.
69. Moir, *Discovery of Britain*, 35–6.
70. Defoe, *Tour Through Great Britain*, 217.
71. J. J. Cartwright (ed.), *The Travels through England of Dr Richard Pococke, successively Bishop of Meath and of Ossory during 1750, 1751, and later years*, 2 vols, Camden Society, 2nd series (1888–89).
72. Cartwright, *Travels of Dr Richard Pococke*, 130.
73. Ibid. 132.
74. S. Shaw, *A Tour in the West of England in 1788* (London, 1789). The quotation is taken from the extract published in Pearse Chope, *Early Tours*, 219.
75. DRO, 564M/F1–17; Royal Institution of Cornwall, DJS/2/1–2; see also Gray, *The*

Garden History of Devon, 245–7; P. L. Hull, 'Four Cornish manuscripts', *Journal of the Royal Institution of Cornwall*, new series, 3 (1957), 32–42. See also, 'A tour in Cornwall in 1780 by the Rev. John Swete' printed with a biographical note by R. Leach in *Journal of the Royal Institution of Cornwall*, new series, 6 (1971), 185–219.

76. Swete was, for example, a contemporary of the agricultural topographers William Marshall and Robert Fraser.
77. W. G. Maton, *Observations relative chiefly to the Natural History, Picturesque Scenery and Antiquities of the Western counties of England, Made in the years 1794 and 1796*, 2 vols (Salisbury, 1797); see also BL, Add. MS 32, 442–3.
78. The quotation is taken from the extract published in Pearse Chope, *Early Tours*, 277–8.
79. Ibid. 236–7.
80. W. Marshall, *The Rural Economy of the West of England*, 2 vols (London, 1796).
81. R. Fraser, *General view of the agriculture of the county of Devon* (London, 1794); J. Billinsgley, *General view of the agriculture of the county of Somerset* (London, 1795).
82. Carew, *Survey of Cornwall*, new edn (London, 1769), 67.
83. Ibid. 11.
84. See Youings, *Sixteenth-Century England*, 357–8; Greenslade, 'Introduction: county history', 13–20; Carew, *Survey of Cornwall*, new edn (London, 1769), 'The life of Richard Carew Esq', xxii–xxiii; J. Youings, 'Devon's first local historians', *Devon Historian*, 1 (1970), 5–8.
85. Youings, 'Devon', 119; Gray, *Garden History of Devon*, 19.
86. T. Risdon, *The Chorographical Description, or survey of the County of Devon, with the City and County of Exeter* (Plymouth, 1811), 95.
87. Carew, Survey of Cornwall, new edn (London, 1769), 21.
88. Risdon, *Chorographical Description* (Plymouth, 1811), 7.
89. Ibid. 8.
90. Carew, *Survey of Cornwall*, new edn (London, 1769), 11.
91. Ibid. 22.
92. Risdon, *Chorographical Description* (Plymouth, 1811), 4–7.
93. Carew, *Survey of Cornwall*, new edn (London, 1769), 24.
94. Ibid. 113.

2 John Leland in the West Country

Abbreviations

Chandler: Chandler, John (ed.) *John Leland's Itinerary: Travels in Tudor England* (Stroud, 1993).

Hearne: Hearne, Thomas (ed.) *J. Lelandi antiquarii de rebus Britannicis Collectanea . . . editio altera*, 6 vols (London, 1774)

Smith: Smith, L.Toulmin (ed.) *The Itinerary of John Leland In or About the Years 1535–1543*, 5 vols (London, 1906–10)

1. For more detail see Chandler, xi–xxxi, and sources there cited.
2. The *Commentarii de Scriptoribus Britannicis*, ed. by A. Hall in 1709; the *Itinerary*, ed. by Thomas Hearne in 1710–12, later editions in 1744–5 and 1768–70; the *Collectanea*, ed. by Thomas Hearne in 1715, later editions in 1770 and 1774.
3. Smith.
4. R. Pearse Chope (ed.) *Early Tours in Devon and Cornwall* (Exeter, 1918), 1–82. Chope's introduction (xi–ii) makes it clear that his work is based on a conflation of texts in Hearne's edition, rather than Smith's; his detailed annotations remain useful.
5. See a bibliography of recent scholarship in Chandler, xxxiv–vi.
6. Chandler.
7. Bodleian, MS Top. Gen. c.3, 253–74; printed in Hearne, vol. 4, 148–68.
8. Smith, vol. 1, 143–4; see J. P. Carley, 'John Leland at Somerset libraries', *Somerset Archaeology and Natural History*, 129 (1987), 141–54. This tour could hardly have begun before June 1533, since Leland was heavily involved in Anne Boleyn's coronation in London at the end of May: Letters and Papers of Henry VIII [hereafter L&P Hen8], vol 6, 251 (564); his contribution is printed in J. Nichols, *The progresses, and public processions, of Queen Elizabeth* (London, 1788), vol. 1, i–xix.
9. J. P. Carley, 'John Leland in Paris: the evidence of his poetry', *Studies in Philology*, 83 (1986), 1–50.
10. Hearne, vol. 4, 150.
11. Ibid. 154: 'Ossa Arturi levata erant ex sacro cimiterio anno domini M. centesimo octuagesimo nono per Henricum Sully, Glasconiae abbatem'. In the manuscript (Bodleian MS Top. Gen. c.3, 261) this sentence is followed by these words, struck through: 'et sepulta a'te ara' sci' Stephani. Postea an'o d'm MCC septuagesimo octavo Edwardo rege cura'te et admone're translata fuere et sepulta ante sum'a' ara'.' *Codrus* is printed in Hearne, vol. 5, 2–10.
12. Smith, vol. 1, 315–26.
13. He was at York in 1534, where he defaced a tablet in the minster: L&P Hen8, vol. 7, 637 (app. 23). He was in London in January 1537 and July 1539: L&P Hen8, vol. 12 (1), 112 (230); vol. 14 (1), 545 (1219). His Welsh journeys are believed to have taken place between 1536 and 1539: Smith, vol. 3, viii–ix.
14. *Naeniae in mortem Thomae Viati equitis incomparabilis* (London, 1542). Translation in K. Muir, *Life and Letters of Sir Thomas Wyatt* (Liverpool, 1963), 261–9. See also P. Thomson, *Sir Thomas Wyatt and his Background* (Stanford, California, 1964).
15. Smith, vol. 1, 107–285.
16. Ibid. 151.
17. *Assertio* is printed in Hearne, vol. 5, 11–64; quotation from p. 29: 'Dii boni, quantum hic profundissimarum fossarum? Quot hic egestae terrae valla? Quae demum praecipitia? Atque, ut paucis finiam, videtur mihi quidem esse et artis et naturae miraculum.'
18. *Genethliacon* is printed in T. Hearne, *The itinerary of John Leland the antiquary . . .*, 2nd edn (1744), vol. 9. Leland's preamble apologises for the delay in completing the work, 'libellus ante aliquot annos inchoatus, et nunc vero absolutus, et editus'.

19. See Chandler, xxv. An exception is Leland's itinerary account of Guy's Cliff, near Warwick, which lapses into enthusiastic poetic Latin.
20. *Cygnea Cantio*, with its commentary, is also printed in Hearne, *Itinerary*.
21. Smith, vol. 5, 71–111.
22. Ibid. vol. 1, xxiii–xxiv; vol. 5, 69, note. Its abrupt beginning and ending suggest that the outer pages of the autograph had become damaged or lost by the time that Stow copied it *c*.1576.
23. Chandler, xxx, where I suggest 1545. This is far from certain, however, and my arguments there are weakened by two considerations: Sir John Poyntz of Alderley, Gloucestershire, seems still to have been alive when Leland visited (Smith, vol. 5, 95), and yet he died on 29 November 1544 (Burke's *Landed Gentry*, 18th edn, 1965, vol. 1, 581); the preface to *Cygnea Cantio* is dated from London 3 Kalends July (i.e. 30 June) 1545. Leland had been in Oxford on 20 May 1545 (L&P Hen8, vol. 20 (1), 388 (776)).
24. *Commentarii de scriptoribus Britannicis* . . ., 1709. Boniface (Winfridus) is described on 126–8. A new edition, with translation, of the *Commentarii* by C. Brett is in preparation.
25. *Encomia illustrium virorum*, printed from the first edition of 1589, in Hearne, vol. 5, 79–167 (Ad Jo. Clericum, Epis. Badonicum, 111–12; De Thermis Britannicis, 90).
26. See, for example Smith, vol. 5, 220–4 and 230–3: quotations from 220, 223 and 230.
27. Ibid. vol. 5, 45; T. Fuller, *The History of the Worthies of England* (London, 1662), 278.
28. Smith, vol. 1, 143–4.
29. Ibid. vol. 1, 248. See A. L. Rowse, *Tudor Cornwall: Portrait of a Society*, 2nd edn, (London, 1969), 247. The inscription survives.
30. Smith, vol. 1, 200. The chart is BL, Cotton MS Aug I i, 35, 36, 38, 39. See A. H. W. Robinson, *Marine Cartography in Britain* (London, 1962), 20.
31. M. Rule, *The Mary Rose: The Excavation and Raising of Henry VIII's Flagship*, 2nd edn (1983), 30–8.
32. Smith, vol. 1, 239–42.
33. Smith's map 2 (in vols 1 and 5) is therefore misleading.
34. Bodleian, MS Top. Gen. e 10, ff. 40–1.
35. See note in Smith, vol. 1, 182. The passages in question are printed on 180–1 and 285–306.
36. G. Edelen (ed.) *The Description of England by William Harrison* (Cornell, 1968), 4 and note 5.
37. See J. H. Harvey (ed.) *William Worcestre itineraries* (Oxford Medieval Texts, 1969).
38. Most recently by M. W. Greenslade in C. R. J. Currie and C. P. Lewis (eds) *English County Histories: A Guide* (Stroud, 1994), 9: 'The father of English local history was John Leland'.
39. Edelen, *Description of England*, 217–18 and note 45. An example of material from a lost East Anglian journey may occur in a description of the River Welland which

Harrison ascribes to Leland: see Holinshed's *Chronicles of England, Scotland and Ireland* (1807), vol. 1, 171.

40. I should like to record my thanks to Veryan Heal and the Devon Archaeological Society, who invited me to lecture on this subject, and to Robert Higham and Mark Brayshay for their encouragement and support.

3 Some Early Topographers of Devon and Cornwall

1. F. E. Halliday (ed.) *Richard Carew of Antony: The Survey of Cornwall* (London, 1953), 103.
2. Ibid. 123.
3. Halliday, *Carew of Antony*.
4. Devon Record Office [DRO], Z19/18/19 John Hooker, 'Synopsis Corographical of the county of Devon', 116.
5. The most easily accessible version is that edited by W. J. Harte, J. W. Schopp and H. Tapley Soper, in *John Vowell, alias Hooker, The Description of the Citie of Excester*, (Devon and Cornwall Record Society, 1919 and 1947), 55–96.
6. DRO, Exeter City Records, Book 51.
7. See Joyce Youings, 'John Hooker and the Tudor bishops of Exeter', in M. Swanton (ed.) *Exeter Cathedral: A Celebration* (1991), 202–7.
8. British Library, Harleian MS 5827; and W. J. Blake, 'Hooker's Synopsis Corographical of Devonshire', *Transactions of the Devonshire Association*, 47 (1915), 334-68.
9. DRO, Z19/18/19.
10. Ibid. 261–2.
11. G. Edelen (ed.) *Description of England by William Harrison* (Cornell, 1968). This was first published in Holished's *Chronicles* in 1577 and revised for the second edition of 1587 of which Hooker was the chief editor. Sir Thomas Smith lifted the same material from Harrison for his *De Republica Anglorum* (1583), so Hooker was in good company.
12. T. Westcote, *View of Devonshire in* MDCXXX, ed. by George Oliver and Pitman Jones (Exeter, 1845), 294–5.
13. Ibid. 131.
14. Ibid. 449.
15. Ibid. 449 and 452.
16. Ibid. 69.
17. DRO, Z19/18/19, 35.
18. Westcote, *Devonshire*.
19. T. Risdon, *Chorographical Description or Survey of the County of Devon*, ed. by E. Curll (London, 1714); *Continuation of the Survey of Devonshire*, ed. by E. Curll (London, 1723–33); and *Chorographical Description of Devonshire*, ed. by Mr Rees *et al.* (London, 1811).

20. W. Pole, *Collections towards a Description of Devon* (London, 1791).
21. A version of this paper, limited to the county of Devon, was printed in C. R. J. Currie and C. P. Lewis (eds) *English County Histories: A Guide* (Stroud, 1994), which also contains a chapter on Cornwall by John Walker.

4 Somerset Topographical Writing, 1600–1900

1. R. Gough, *British Topography*, (1780) vol. 1, x. Dr Jonathan Barry read an earlier version of this paper and offered some valuable suggestions and references unknown to me.
2. *Somerset and Dorset Notes and Queries*, 5 (1897), 77–102. Gerard's Dorset survey was published in 1732.
3. John Batten, the 'discoverer' of Gerard, saw the volume when its owner, the earl of Winchelsea, deposited it for examination with the Historical Manuscripts Commission. It is now in the Northamptonshire Record Office (FH 113). It was edited by the Revd E. H. Bates for the Somerset Record Society: *The Particular Description of the County of Somerset*, (Somerset Record Society xv, 1900).
4. Bates, *Particular Description*, 9.
5. Ibid. 11.
6. Ibid. 131–2.
7. Ibid. xx.
8. D. C. Douglas, *English Scholars 1660–1730*, 2nd edn (London, 1951), 32.
9. Bates, *Particular Description*, xx.
10. Vicar of Frome 1662–72; rector of Bath 1666–80; rector of Street 1672–80.
11. Rector of Sock Dennis 1639–83; vicar of Yeovil 1660–83.
12. Articles on all three in *Dictionary of National Biography*; see also Gough, *Topography*, vol. 2, 221, 223.
13. Gough, *Topography*, vol. 1, 187.
14. Ibid. 189.
15. Ibid. 187.
16. J. Foster (ed.) *Alumni Oxonienses 1500–1714* (Oxford, 1892).
17. Bodleian, MS Aubrey 13, f. 59 and onwards, and MS 15.
18. A. J. Turner, 'Andrew Paschall's Tables of Plants for the Universal Language', *Bodleian Library Record*, 9 (1978), 346–50.
19. *Somerset and Dorset Notes and Queries*, 27 (1954–60), 93.
20. Gough, *Topography*, vol. 2, 222, 225.
21. British Library, Additional, MS 4162; Lambeth Palace Library, MS 942/34; printed in S. Heywood, *Vindication of Mr Fox's History* (1811). The shorter account, found in the vaults of Hoare's Bank, is printed in *Somerset and Dorset Notes and Queries*, 28 (1961–7), 15–21.
22. *Remarks and Collections of Thomas Hearne* (hereafter *Hearne's Collections*), vol. 8 (Oxford Historical Society, 1907), 328, 334, 373.

23. R. Rawlinson (ed.) *Miscellanies on Several Curious Subjects* (London, 1714).
24. S. W. Rawlins and P. B. G. Binnall, 'Dr George Harbin', *Somerset Archaeological and Natural History Society Proceedings*, 93 (1947), 68–83.
25. *Hearne's Collections*, vol. 2, (Oxford Historical Society, 1886), 342.
26. Ibid. vol. 5 (Oxford Historical Society xlii, 1901), 75, 80–1, 109, 150, 152, 157, 166, 173–4, 176–8, 180, 184-5, 225, 235, 353; ibid. vol. 6 (Oxford Historical Society xliii, 1902), 42, 45.
27. Ibid. 133; ibid. vol. 7 (Oxford Historical Society xlviii, 1906), 224.
28. Ibid. 172, 333, 387.
29. Ibid. vol. 8 (Oxford Historical Society l, 1907), 15, 55, 65–6.
30. Ibid. vol. 10 (Oxford Historical Society lxvii, 1915), 364, 370.
31. Ibid. 420.
32. Somerset Record Office [SRO], DD/HN.
33. *Hearne's Collections*, vol. 11 (Oxford Historical Society lxxii, 1921), 10.
34. SRO, DD/TB 16/18.
35. Rawlins and Binnall, 'George Harbin', 81.
36. SRO, DD/SH 96.
37. S. Keynes, 'George Harbin's transcript of the lost cartulary of Athelney Abbey', *Somerset Archaeology and Natural History Society Proceedings*, 136 (1992), 149–59.
38. SRO, DD/TB 19/1–3, 20/1–5, 21/1–2, 51/1.
39. Ibid. 17/12, 20/1.
40. *Hearne's Collections*, vol. 5, 13, 16.
41. Ibid. 256.
42. Ibid. vol. 11, 46.
43. Ibid. 237.
44. Information from Dr Simon Keynes.
45. SRO, DD/AH 21/1–2, 60/10.
46. *Hearne's Collections*, vol. 7, 236.
47. Le Neve, *Fasti Ecclesiae Anglicanae, 1541-1857, Bath and Wells*, ed. by J. M. Horn and D. S. Bailey, 77, 93; Foster, *Alumni Oxonienses*.
48. *Hearne's Collections*, vol. 6, 283.
49. *Hearne's Collections*, vol. 7, 236.
50. Information from Dr Jonathan Barry; see also Bodleian, MS Willis, 43, 46, 60, 64.
51. The original volume was offered for private sale in 1993 and its present location is not known. A photocopy is held: SRO, DD/X/WBB 49.
52. M. McGarvie, 'The antiquities of Mells, Elm and Buckland in 1730', *Transactions of the Ancient Monuments Society* 27 (1983), 77–104.
53. V. A. Eyles, 'Scientific activity in the Bristol region in the past', in C. M. MacInnes and W. F. Whittard (eds) *Bristol and its Adjoining Counties* (British Association, 1955), 131.
54. *Hearne's Collections*, vol. 10, 254.
55. Ibid. 256.
56. *Hearne's Collections*, vol. 11, 46.

57. *Hearne's Collections*, vol. 10, 426.
58. SRO, DD/SH 204, with the bookplate of Sir William Blackstone.
59. SRO, DD/SH 192.
60. Ibid. 194, 200.
61. Ibid. 196.
62. Ibid. 141, 251–5, 257–70, 272.
63. *Hearne's Collections*, vol. 10, 426; vol. 11, 46.
64. Ibid. vol. 11, 46.
65. T. Chubb, A *Descriptive List of the Printed Maps of Somersetshire, 1575–1914* (1914), 35. Thomas Ford sent several corrections: SRO, DD/SH 41–6.
66. SRO, DD/SH 110.
67. Ibid. 92.
68. Ibid. 103.
69. Ibid. 107–8.
70. *Somerset County Herald*, 15 June 1940.
71. T. Cox, *Magna Britannia et Hibernia antiqua et nova . . .*, part 60 (1727), 790–912.
72. For detailed references to Collinson see R. W. Dunning, 'Introduction', in J. Collinson, *The History and Antiquities of the County of Somerset* (Microprint edition, Gloucester, 1983).
73. E. Green, *Bibliotheca Somersetensis* (1905), vol. 2, 317.
74. Bristol Record Office, Ashton Court MSS (AC), uncatalogued boxes; photocopies in SRO.
75. A text of the survey is planned by the Somerset Record Society.
76. H. C. Maxwell Lyte, *A History of Dunster* (1909), vol. 1, vi.
77. J. Rutter, *Delineations of the North Western Division of the County of Somerset and of its Antediluvian Bone Caves* (Shaftesbury, 1829).
78. Vicar of Bicknoller 1811–54; vicar of Meare 1824–56; W. Phelps, *Calendarium Botanicum* (1810).
79. BL, Add. MSS 33820–3, 33830.
80. W. Phelps, *The History and Antiquities of Somersetshire*, 2 vols in 8 parts (1836–9); *Somerset and Dorset Notes and Queries*, 32 (1988), 644–55.
81. SRO, DD/SAS SI 5; DD/DN 293.
82. *Somerset Archaeological and Natural History Society Proceedings*, 1 (1851), 5, 50. For the text of the questionnaires: 75–85.
83. Ibid. 9 (1859), 2–5.
84. Ibid. 10 (1860), 2.
85. Ibid. 9 (1859), 4; 10 (1860), 3–4.
86. F. W. Weaver and E. H. Bates with J. R. Bramble (eds) *Index to Collinson's History of Somerset* (1898).
87. *Somerset Archaeological and Natural History Society Proceedings*, 36 (1890), 17–23, 35; 44 (1898), 8; 45 (1899), 10–11.
88. R. B. Pugh (ed.) *Victoria County History* General Introduction (1970), 1.
89. Institute of Historical Research, VCH correspondence, Dorset.

90. Taunton, County Hall, VCH correspondence.
91. *Victoria County History: Somerset*, vol. 1 (1906), xiii.
92. *Two Cartularies of . . . Muchelney and Athelney* (Somerset Record Society xiv, 1899).
93. Bates, *Particular Description of Somerset.*
94. *Somerset Archaeological and Natural History Society Proceedings*, 64 (1909), 92.

5 Early Topographers, Antiquarians and Travellers in Dorset

1. L. Toulmin Smith, *The Itinerary of John Leland*, vols 1, 4 and 5 (London, 1906–10). See also J. Chandler, *John Leland's Itinerary: Travels in Tudor England* (Stroud, 1993).
2. R. Legg (ed.) *Thomas Gerard's General Description of Dorset* (1980); Revd E. H. Bates (ed.) *The Particular Description of the County of Somerset* (Somerset Record Society xv, 1900).
3. Legg, *General Description*, introduction.
4. Bates, *Particular Description*, introduction. See also A. Sandison, *Trent in Dorset* (Dorchester, 1969), 35–42.
5. Extracts from *Thomas Gerard's General Description of Dorset* are taken from R. Legg's edition of 1980.
6. P. Mundy, *A Petty Progress through England and Wales, and his Tour round the Coast*, Hakluyt Society, 2nd series, iv (1925). See also N. M. Richardson, 'The travels of Peter Mundy in Dorset in 1635', *Dorset Natural History and Archaeological Society Proceedings*, 42 (1922), 42–50.
7. L. G. Wickham Legg (ed.) *A Relation of a Short Survey of the Western Counties made by a Lieutenant of the Military Company in Norwich in 1635*, Camden Miscellany, XVI (1936).
8. T. Fuller, *The Worthies of England*, ed. by J. Freeman (London, 1952).
9. C. Morris (ed.) *The Journeys of Celia Fiennes* (London, 1947), 10–13.
10. Daniel Defoe, *A Tour through England and Wales* (Everyman Edition, 1927), vol. 1, 206–18.
11. J. J. Cartwright (ed.) *The Travels through England of Dr Richard Pococke, successively Bishop of Meath and of Ossory during 1750, 1751, and later years*, Camden Society, new series, XLIV, (II) (1888–9), 44–5, 136–50.
12. W. Stukeley, *Itinerarium Curiosum* (London, 1724); A. M. Broadley, *The Visit of Bro. Dr W. Stukeley to Dorchester* (Weymouth, 1913); S. Piggott, *William Stukeley: An Eighteenth-Century Antiquary* (Oxford, 1950), 67–74.
13. Arthur Young, *A Farmer's Tour through the East of England*, vol. 3 (1771), 245–411.
14. A full list of topographical and historical works on Dorset can be found in Robert Douch's invaluable bibliography, *A Handbook of Local History: Dorset*, 2nd edn (1962).

15. A full account of the life of John Hutchins and an assessment of the value of his county history can be found in the introduction to the reprint of the 3rd edition of the work edited by Robert Douch in 1973. This introduction is also printed in J. Simmons (ed.) *English County Historians* (Wakefield, 1978), 113–58. See also J. H. Bettey, 'Dorset', in C. R. J. Currie and C. P. Lewis (eds) *English County Histories: A Guide* (Stroud, 1994), 125–31.

6 From Romanticisim to Archaeology: Richard Colt Hoare, Samuel Lysons and Antiquity

1. S. Piggott, *Ruins in a Landscape* (Oxford, 1975), 18–21.
2. S. Piggott, 'Prehistory and the Romantic Movement', *Antiquity*, 11 (1937), 31–8.
3. K. Woodbridge, *Landscape and Antiquity* (Oxford, 1970), 187–217.
4. *Quarterly Review*, 10 (1811), 111–20; 11 (1811), 440–8.
5. W. G. von Goethe, *Tagebuch*, 18 June 1816.
6. *Gentleman's Magazine*, 81 (1811), 418–22.
7. J. Evans, *A History of the Society of Antiquaries* (Oxford, 1956), 204; 219–20. Thomas Lawrence's portrait of Lysons appears as Pl. 34.
8. S. Lysons, *An Account of the Remains of a Roman Villa discovered in the County of Sussex* (London, 1815).
9. D. Lysons and S. Lysons, *Magna Britannia, being a concise topographical account of the several counties of Great Britain, volume the sixth containing Devonshire* (London, 1822), cccvi.
10. Ibid. cccxxii.
11. *Antiquaries Journal*, 64 (1984), 261–5.
12. *Proceedings of the Devon Archaeological Society*, 33 (1975), 193–5.
13. T. Rickman, *An Attempt to Discriminate the Styles of English Architecture from the Conquest to the Reformation* (London, 1819).

7 The Scientific Gaze: Agricultural Improvers and the Topographical of South-West England

1. S. Wilmot, *'The Business of Improvement': Agriculture and scientific culture, c.1770–1870* (Bristol, 1990).
2. Sir E. J. Russell, *A History of Agricultural Science in Great Britain, 1620–1954* (London, 1966), 22.
3. W. Marshall, *Minutes of Agriculture* (London, 1778), 10–11, 120–2, 149; Wilmot, *Business of Improvement*, 2–20.
4. R. C. Allen and C. O'Grada, 'On the road again with Arthur Young: English, Irish and French agriculture during the Industrial Revolution', *Journal of*

Economic History 48 (1988), 93–116; E. Kerridge, 'Arthur Young and William Marshall', *History Studies* 1 (1968) 43–53.

5. Allen and O'Grada, 'On the road again with Arthur Young', 94, 98–100.
6. W. Marshall, *Minutes, Experiments, Observations and General Remarks on Agriculture in the Southern Counties*, 2 vols (London, 1799), vol. 2, 116, 118, 122–3, 311–12.
7. W. Marshall, *The Rural Economy of the West of England*, 2 vols (London, 1796), vol. 1, 280–1.
8. J. Barrell, *The Idea of Landscape and the Sense of Place, 1730–1840* (Cambridge, 1972), 89–91.
9. J. Billingsley, *General View of the Agriculture of the County of Somerset* (London, 1795), i–ii.
10. Sir E. Clarke, 'The Board of Agriculture, 1793–1822', *Journal of the Royal Agricultural Society of England*, 59 (1898), 1–42, esp. 16.
11. Marshall, *Rural Economy*.
12. R. J. P. Kain and H. C. Prince, *The Tithe Surveys of England and Wales* (London, 1985), 109–112. The original records are held at the Public Record Office [PRO], IR/18 Tithe files.
13. W. F. Karkeek, 'On the farming of Cornwall', *Journal of the Royal Agricultural Society of England*, 6 (1845), 400–62; H. Tanner, 'The farming of Devonshire', *Journal of the Royal Agricultural Society of England*, 9 (1849) 454–95; F. Punchard, 'Farming in Devon and Cornwall', *Journal of the Royal Agricultural Society of England*, 51 (1890), 511–36; the prize essays on Somerset awarded in 1850 were re-published in 1851 as T. D. Acland and W. Sturge, *The Farming of Somersetshire*, (London, 1851). References to Acland and Sturge, *Farming in Somersetshire* in this chapter refer to the 1851 publication; L. H. Ruegg, 'Farming of Dorsetshire', *Journal of the Royal Agricultural Society of England*, 15 (1854), 389–454.
14. Kerridge, 'Arthur Young and William Marshall', 46; P. Horn, *William Marshall (1745–1818) and the Georgian Countryside* (Abingdon, 1982).
15. W. Marshall, *Review and Abstract of the County Reports to the Board of Agriculture*, 5 vols (London, 1818). Vol. 5 includes the counties of South-West England.
16. Allen and O'Grada, 'On the road again with Arthur Young', 115.
17. Marshall, *Review and Abstract*, vol. 5, 550–1, 515.
18. Ibid. 527–8.
19. Ibid. 557, 559–62.
20. Kerridge, 'Arthur Young and William Marshall', 50–1.
21. Ibid. 51.
22. Horn, *William Marshall*, 27–31.
23. Biographical bibliographies are listed in Wilmot, *Business of Improvement*, 100; coverage of the nineteenth-century agriculturalists in national bibliographies is poor and local reference sources may well be more helpful sources of information.

24. A. H. D. Acland, *Memoirs and Letters of the Right Honourable Sir Thomas Dyke Acland* (London, 1902), 142, 153.
25. H. C. Prince, 'The changing rural landscape, 1750–1850', in G. E. Mingay (ed.) *The Agrarian History of England and Wales*, vol. 6 (Cambridge, 1989), 10, 77–8.
26. G. E. Mingay (ed.) *The Victorian Countryside*, 2 vols (London, 1981); J. D. Chambers and G. E. Mingay, *The Agricultural Revolution, 1750–1880* (London, 1966).
27. A. Young, *Farmers' letters to the people of Great Britain*, 3rd edn (London, 1771), 292.
28. J. Beasley, *A lecture delivered to the members of the Faringdon Agricultural Book Club* (London, 1860), 7.
29. Marshall, *Rural Economy*, vol. 1, xxviii; A. Briggs, *The Age of Improvement, 1783–1867* (London, 1959), 34.
30. *British Parliamentary Papers*, 'First report to the Select Committee appointed to take into consideration the improvement of waste, uninclosed and unproductive lands of the Kingdom', presented by Sir John Sinclair to the House of Commons, Dec. 1795, 15.
31. J. C. Loudon, *An encyclopaedia of Agriculture* (London, 1835), 207.
32. J. Wrightson, *The principles of agriculture as an instructional subject* (London, 1888), 208–9; H. J. Webb, *Advanced Agriculture* (London, 1894), 398–9.
33. A. Young, *Farmers' letters to the landlords of Great Britain*, 3rd edn (London, 1771), 108, 298; J. L. Morton, *The resources of estates* (London, 1858), concluding remarks to chapter 1, 1–26.
34. Loudon, *Encyclopaedia*, 752.
35. D. G. F. Macdonald, *Hints on Farming and Estate Management*, 5th edn (London, 1865), 40.
36. Morton, *Resources of estates*, conclusion.
37. J. J. Mechi, *How to farm profitably—or the sayings and doings of Mr Alderman Mechi* (London, 1859).
38. J. Grant, 'A few remarks on the large hedges and small enclosures of Devonshire and the adjoining counties', *Journal of the Royal Agricultural Society of England*, 5 (1844), 420–9, esp. 420.
39. Grant, 'Remarks on large hedges', 424.
40. Ibid. 422.
41. Ibid. 429.
42. Macdonald, *Hints on farming*, 56–7, 62. Views on the agricultural undesirability of hedges did not change in the twentieth century: see A. D. Hall, *A Pilgrimage of British Farming, 1910–12* (London, 1913), 357, for the following comment: 'When the old landlords drained and marled to improve their estates, it is a pity they did not also re-map them'.
43. Sturge, 'Farming of Somersetshire', 169.
44. Karkeek, 'Farming of Cornwall', 402.
45. PRO, IR/18, Tithe files for East Buckland, Devon.

46. C. Vancouver, *General view of the agriculture of Devon* (London, 1808), 217.
47. Tanner, 'Farming of Devonshire', 485.
48. PRO, IR/18, Tithe files for High Bray and Challacombe, Devon.
49. PRO, IR/18, Tithe files for Monksilver, Stoke Pero, Brushford, Withypool, Dulverton, and Hawkridge, for example.
50. S. Wilmot, 'Agrarian systems in South-west England in the nineteenth century', unpubl. paper presented at fifth annual symposium of the Centre for South-Western Historical Studies, Exeter, November 1990, 5.
51. PRO, IR/18, Tithe files for St Giles-on-the-Heath and Stoke Rivers, Devon.
52. Marshall, *Rural Economy*, vol. 1, xxxii–xxxiii.
53. Ibid. 130–33, 141, 206–7.
54. J. Claridge, *General view of the agriculture in the county of Dorset* (London, 1793), 5–6.
55. Ruegg, 'Farming of Dorsetshire', 437.
56. *British Parliamentary Papers*, Reports of Assistant Commissioners to the Royal Commission on Agriculture, 1882. W. C. Little's Fourth Report, 32.
57. Ruegg, 'Farming of Dorsetshire', 400, 414, 439.
58. Ibid. 440, 448.
59. W. Stevenson, *General view of the agriculture of the county of Dorset* (London, 1812), 91, 460.
60. Little, Fourth Report, 35.
61. Ruegg, 'Farming of Dorsetshire', 417, 453; Little, Fourth Report, 34 (my italics).
62. Tanner, 'Farming of Devonshire', 461, 466, 493.
63. Ibid. 484.
64. H. S. A. Fox, 'The functioning of bocage landscapes in Devon and Cornwall', in *Les Bocages: Histoire, Ecologie, Economie*, (Centre National de la Recherche Scientifique, Rennes, 1976), 55–61, esp. 56–7, with grateful acknowledgements to Harold Fox for kindly sending the author a copy of this paper.
65. T. D. Acland, 'On the farming of Somersetshire', *Journal of the Royal Agricultural Society of England*, 11 (1850), 696.
66. R. Dymond, 'Devonshire fields and hedges', *Journal of the Bath and West of England Society*, new series 4 (1856), 132–48, esp. 133–6. The author is grateful for the kindness of Anthony Collings in sending a copy of this paper.
67. Hall, *Pilgrimage of British Farming*, 356–7.
68. S. Wilmot, 'Landownership, farm structure and agrarian change in South-West England: Regional experience and national ideals' (unpubl. PhD thesis, University of Exeter, 1988), 109.
69. L. D. Stamp, 'Devonshire', in L. D. Stamp (ed.) *The Land of Britain* (London, 1941), part 92, 471–544, esp. 518–9.
70. T. R. B. Dicks, 'The south-western peninsulas of England and Wales: studies in agricultural geography, 1550–1900' (unpubl. PhD thesis, University of Wales, 1964), 76.
71. Wilmot, 'Agrarian systems in South-west England', 8.
72. Punchard, 'Farming in Devon and Cornwall', 522; W. Harwood Long, 'Factors

affecting some types of farming in Devon and Cornwall', *Journal of the Royal Agricultural Society of England*, 94 (1933), 42–61.

73. *British Parliamentary Papers*, Reports of Assistant Commissioners to the Royal Commission on Agriculture, 1882. W. C. Little's Report on Cornwall, 12, 15.
74. Sturge, 'Farming of Somersetshire', 167.
75. Wilmot, 'Landownership, farm structure and agrarian change in South-West England', 109.
76. M. Williams, *The Draining of the Somerset Levels* (Cambridge, 1970), 187–96, 197–227.
77. P. Horn gives a full list of Marshall's publications in *William Marshall*.
78. M. Drabble, *A Writer's Britain* (London, 1979), 68, 130.
79. R. Fraser, *General view of the agriculture of the county of Devon* (London, 1794), 9.
80. Claridge, 'General view of Dorset', 46; Stevenson, 'General view of Dorset', 22; D. Allen, *The Naturalist in Britain* (London, 1976), 70, 125–32.
81. G. Sheldon, *From Trackway to Turnpike* (London, 1928), 117–20.
82. H. Rider Haggard, *Rural England*, 2 vols (London, 1902), 230.
83. Much has been written on changes in landscape tastes, amongst the most useful summaries available are: J. Hayes, *The Landscape Painting of Thomas Gainsborough* (London, 1982); A. Bermingham, *Landscape and Ideology: The English Rustic Tradition, 1740–1860* (Berkeley, 1986); S. Daniels, 'The political iconography of woodland in later Georgian England', in D. Cosgrove and S. Daniels (eds) *The Iconography of Landscape* (Cambridge, 1988), 43–82; P. Howard, 'Painters' preferred places', *Journal of Historical Geography*, 11 (1985), 138–54; K. Thomas, *Man and the Natural World* (London, 1983); C. Payne, *Toil and Plenty: Images of the Agricultural Landscape in England, 1780–1890* (Newhaven and London, 1993).
84. Hayes, *Landscape Painting*; Drabble, *Writer's Britain*, 122–3; Thomas, *Man and the Natural World*, 260, 266.
85. Quoted in Thomas, *Man and the Natural World*, 285.
86. H. Prince, 'Art and agrarian change, 1710–1815', in Cosgrove and Daniels, *Iconography of Landscape*, 98–118, 98–9.
87. J. Barrell, *The Idea of Landscape and the Sense of Place*, 1730–1840 (Cambridge, 1972), chapters 1–2.
88. G. B. Worgan, *General view of the agriculture of Cornwall* (London, 1811), 6 (my italics).
89. Ibid. 7, 104.
90. Marshall, *Rural Economy*, vol. 1, 40–2, 276–7, 280–1, 290; vol. 2, 4, 44–6, 54–5, 59, 71, 110.
91. Ibid. vol. 1, 41; vol. 2, 22.
92. Vancouver, *General view of Devon*, 17–18, 21, 24, 28. The quotation is taken from p. 17.
93. Ibid. 28, 278, 297–8.

94. Ibid. 230.
95. PRO, IR/18 Tithe files for Oare and Stoke Pero, Somerset.
96. Acland, *Farming of Somersetshire*, 33.
97. Sturge, *Farming of Somersetshire*, 159.
98. K. Hudson, *The Bath and West: A Bicentenary History* (Bradford-on-Avon, 1976), 103–8.
99. T. D. Acland, 'Report on the arts department of the exhibition at Barnstaple', *Journal of the Bath and West*, 8 (1860), 45.
100. Barrell, *Idea of Landscape*, chapter 2, esp. p. 73.
101. Marshall, *Rural Economy*, vol. 1, 289.
102. Loudon, *Encyclopaedia*, 688, 751.
103. Acland, *Farming of Somersetshire*, 95.
104. Grant, 'Remarks on large hedges', 426.
105. Ibid. 426 footnote.
106. Daniels, 'Political iconography of woodland', 62; Thomas, *Man and the Natural World*, 266.
107. Worgan, *General view of Cornwall*, 98, 101, 104.
108. Thomas, *Man and the Natural World*, 286.
109. Ruegg, 'Farming of Dorsetshire', 410, 453.
110. Wilmot, 'Agrarian systems in South-west England', 15–16.
111. J. Caird, *The Landed Interest and the Supply of Food* (London, 1878), 143, (my italics).

INDEX